SMALL FARM
BIG DREAMS

Turning a flower-growing passion into a successful floral business

JENNIFER + ADAM O'NEAL

Publisher: BLOOM Imprint

Authors: JENNIFER + ADAM O'NEAL

Photographers: JENNIFER + ADAM O'NEAL

Editorial Director: DEBRA PRINZING

Creative Director: ROBIN AVNI

Cover Design: JENNY MOORE-DIAZ

Image Editor: HEATHER MARINO

Copy Editor: BRENDA SILVA

Cover + Title Page Photography: ALY CARROLL PHOTOGRAPHY

Copyright © 2022 by BLOOM Imprint.
All rights reserved. No part of this publication may be reproduced, stored in a retrieval system or transmitted, in any form or by any means, electronic, mechanical, photocopying, recording or otherwise, without prior written permission of the publisher.

ISBN: 978-1-7368481-4-2

Library of Congress Control Number: 2022933539

BLOOM Imprint
4810 Pt. Fosdick Drive NW, #297, Gig Harbor, WA   98335
www.bloomimprint.com

Printed in the U.S.A. by Consolidated Press, Seattle, WA

*In loving memory of Sande, our Mom,
a lover of flowers and secret gardens*

SMALL FARM, BIG DREAMS

8
INTRODUCTION

15
CHAPTER ONE
Knowledge Building
research + tips
garden design basics

74
CHAPTER FOUR
Tie a Ribbon
cutting your blooms
garden-inspired design
weddings + events

36
CHAPTER TWO
Choosing the Flowers
10 easiest cut flowers
perennials vs. annuals
design-driven flower farming

98
CHAPTER FIVE
Off to Market
farmers' market | CSA
you-pick | other opportunities
children in the garden

60
CHAPTER THREE
Growing the Flowers
starting from seed
ways to water

114
CHAPTER SIX
The Dormant Season
winter preparation
dahlias are different
prepping + planning
creative time

127
EPILOGUE

INTRODUCTION

It started with a dream.

LIKE ALL DREAMS, OURS FELT LIKE IT WAS AT OUR FINGERTIPS, BUT ALWAYS slightly out of reach. Owning a farm and tending to a piece of land, growing flowers and living with its joy, excitement, and endless possibilities felt like something we'd never have for ourselves — even though it's exactly what we dreamed about nearly each and every day.

For over 10 years, we held onto our little dream of owning a farm. We kept the dream close to our hearts, unknowingly keeping our passion alive as well, through both practice and patience. Looking back, it's easy to see we were on the right path and working towards our goal, whether we knew it or not at the time.

We grew and nurtured a pretty little backyard garden plot at the first home we purchased together in Nashville, Tennessee. The garden contained veggies and cut flowers, with the highlight being a wild pollinator garden full of fragrant bee balm, as well as other pollinator-friendly flowers. Hours were spent each spring at local nurseries picking out new flower varieties to add to our personal garden oasis. It didn't take us long to start collecting exciting specimens, and become self-professed flower addicts in the process.

With our existing experience and interest in gardening, our passion for plants and flowers continued to grow. Family outings became excursions to local greenhouses, where we'd stroll with the kids through the warm structures with the scent of plants as food for our souls. Of course, we'd return home with a plant or two after each visit, even if we didn't have any space left in the garden. During those days, we also spent time visiting local farms, snapping up lots of fresh-flower bouquets at the farmers' market and touring botanical gardens, ogling over plants.

When we finally bought our dream farm, we started small, but like every new landowner, we dabbled in the soil, poking our fingers (and toes) into a little bit of it all. We did everything from raising farm animals and keeping bees, to planting orchards, and growing and canning our own food. But most importantly during our early homesteading days, we grew a small patch of flowers, which we cut and took to our local farmers' market each week.

INTRODUCTION

A fresh-cut bounty of sunflowers puts a smile on Adam's face. We often use our vintage Toyota Matrix, which Adam calls our "Ma-Trux" to transport crops from the large fields in the production area.

Imagine our surprise to witness the happiness our flowers brought to people! It was just as fulfilling to see people enjoy taking photos of our blooms, and we also loved engaging with those who stopped just to smell the flowers. Sharing in and spreading happiness throughout our community with the beautiful flowers we grew was intoxicating and completely contagious. Naturally, we had to grow more!

IT'S ASTOUNDING TO REALIZE THAT WHAT STARTED AS A HUMBLE 20-BY-30-FOOT patch during our first year has grown into a blooming 7.5 acres of cut flowers. In the past 10 years, we've grown our small farm into something BIG that we never thought was possible: working for ourselves, sustaining our family financially, and living our dream each and every day.

PepperHarrow Farm, our home-based enterprise, now grows enough blooms to supply farmers' markets, grocery stores, local florists, event designers, weddings and events, CSAs, as well as customer-cut flower bouquets each week. Over 500 bouquets and a countless bounty of stems leave the farm every week during our growing season.

We've hosted hundreds of guests at the farm through the years for events and workshops, which range from flower farming to floral design. Flower-related classes teach students how to make lavender soap, candles, and other botanical crafts. Then there are the larger-scale events, like our *al fresco* dinner, which is held when the

INTRODUCTION

After a storm passes over the farm, Jenn gathers lisianthus in the cool of dusk. Lisianthus is one of our most popular offerings, equally loved by farmers' market customers and wedding clients.

lavender fields are in full bloom. We love sharing our gorgeous farm with others through these events. It's the time when we can sit back and appreciate our hard work.

WITH GROWING INTEREST IN GETTING BACK TO THE LAND, WE'VE HEARD IT LOUD and clear from our customers and cheerleaders that they, too, want to learn to grow flowers. It's with that encouragement we began to actively share our knowledge about how to grow gorgeous blooms — including more about our flower-farming lifestyle. We've filmed flower-styling tutorials and posted them to our popular YouTube channel, and added more resources on our website, and now, in this very book.

Throughout the pages of "Small Farm, Big Dreams," you'll read our stories of living and raising our family on a flower farm, as well as practical tips and advice to help you become successful at growing flowers. Whether you're new to growing flowers, or a seasoned flower farmer, our hope is that you'll feel our genuine excitement and encouragement, and that the information contained in this book helps you on your journey, regardless of your skill level.

WHETHER YOU'RE GROWING ON A BALCONY, IN THE BACKYARD, OR ON LARGER acreage, our desire is that this book will inspire each of you to dream big. We all have to start somewhere, but the key is to start and keep focusing on your goals. Imagine yourself already there, on your farm, working the soil and breathing in the sweet air.

CHAPTER ONE

Knowledge Building

YOU'RE NOT ALONE IF YOU FEEL LIKE A PAINTER STARING AT A BLANK CANVAS when deciding where you want to start growing flowers. It can be pretty intimidating just choosing which flowers to grow and how to grow them. We can relate to that feeling because we felt the same way 10 years ago. As we looked across our 20 acres of land, we asked ourselves: How much can we handle growing with just the two of us? Will anyone buy our flowers? Which flowers should we grow to be successful, and/or do we want to provide flowers for weddings to local florists?

> You don't need acres of land to be able to grow a beautiful bounty of blooms.

To be totally transparent, you don't need acres of land to be able to grow a beautiful bounty of blooms. Whether you have a small .25-acre backyard or 20 acres and more, we'll share how to maximize what you're able to grow on the space you have available. We will break down those once-overwhelming tasks into manageable steps that feel achievable as you build your confidence.

The key to getting off to a great start with growing flowers is to begin with the basics: put in a lot of research and planning time. It certainly isn't the most glamorous part of flower growing, and it can be hard to find the time to do this — especially when you're eager to start living your dream — but it's incredibly important to pause and do some initial homework.

Our dreaming period took over 10 years before we bought what became PepperHarrow Farm, but we started to visualize it into reality by writing, drawing, and adding clippings into a dream journal, usually on a monthly basis. There were lots of magazine articles about permaculture, flower growing, and other homesteading topics. There were also inventories of flowers we loved and wanted to grow, all collected from seed catalogs we'd received each spring — we even had a running list of farm names we'd brainstormed along the way.

After we bought our farm, we walked our land almost every night for months during that first year, all before we (finally) started breaking ground and planting. This wasn't an easy task — especially since the majority of our land was in pasture, lorded over by four randy horses. Their leader was a huge, handsome buckskin stallion named Tex, who seemed to delight in scaring us on our walks. He ran full bore at our young family, including our then-baby Quinlan, toddler Lochlann, and small son Griffin. He must have smelled the "city-slicker" all over us. We have no idea what that smells like, but we think it might smell a little like a new car!

Aside from the crazy horses, the grass in the pasture was incredibly tall, around 3-4 feet in most areas. We waded through the tall grass until we walked it enough to trample it down a bit. We wore boots, jeans, and long-sleeved shirts to keep the grass from scratching us, and learned to tick-check every time we got back from the walks.

Even though the walks may sound as though they were daunting, they were actually really amazing opportunities to learn and observe our land.

Not only did we take time to appreciate the property we'd purchased, but we also observed the best potential places for growing flowers. Those observations helped us to grow our farm purposefully and with intention in every move and decision. Our desire for success required us to be marathoners, not sprinters.

We saw the places that held too much water and also the places that were prone to wash out when it rained. Getting acquainted with our land in this way rewarded us with priceless knowledge for our future flower gardens — even though, admittedly, we were totally impatient to start growing flowers on Day One.

Our message to all of you is to go slow, be intentional, and do all of the research up front. Doing this will yield great rewards and success as you grow.

Let's Get Started

Everyone's knowledge is just a little bit different when it comes to growing flowers. Some are beginners who have never planted a single seed, and others are semi-pros who just need to know a few more tips and tricks before hitting the mastery level.

Just remember this: no one has 100% of the information they need on Day One, and you're probably going to make some mistakes.

We've made many mistakes over the last 10 years and, even now, still make them occasionally. Keeping a resilient, hard-working spirit will take you far in your flower-growing adventure. Allow every mistake to be good learning lessons, and keep in mind it's often necessary to adapt and shift. One thing this farm has taught us: the mistakes make you learn the right way, and only increase your level of professionalism.

Is there a flower that doesn't grow well for you? Maybe it needs to be taken out of the crop mix. We discovered a few plants that don't do well for us and they were some of the first we took out of our lineup. Pay attention to those things; It's not that you can't or won't come back to them later.

During the mid-winter months, we devote necessary time to planning for the rest of the year. We comb through seed catalogs, map out garden plans, and prepare seed planting schedules.

Research Tips + Tricks

START A DREAM BOOK TO VISUALIZE, document, and take notes on all the things you want to come to fruition or explore.

WATCH ALL OF THE YOUTUBE VIDEOS you possibly can about how to grow flowers and grow them successfully.
If there's a flower that you're interested in growing, focus your research.

FLIP THROUGH SEED CATALOGS and document seed varieties that interest you, but make sure they can grow in your zone and in the cultural conditions of your space (sun/shade).

CHECK OUT YOUR LOCAL PUBLIC television station and watch gardening shows. We started by watching "Victory Garden," "Barefoot Farmer," and "Volunteer Gardener," while we lived in Tennessee.

WORKSHOPS, ONLINE TRAINING, and learning modules with established flower farmers can also be really helpful. We highly recommend any that are onsite, because you really have a chance to get a behind-the-scenes look at how other farms operate. You'll also have a huge opportunity to connect with other growers who are just starting out, which is invaluable, and can often create lifelong connections. Joining Facebook flower farming and gardening groups can be an invaluable way to gain information and grow a peer network.

REACH OUT TO YOUR LOCAL AGRICULTURE EXTENSION OFFICE and see if they offer free University Extension classes on growing. Many do, or they may offer a Master Gardener Program, where you will receive a lot of information and hands-on experience with growing.

READ GARDENING OR FLOWER-FARMING BOOKS! Our favorite is "The Flower Farmer: An Organic Grower's Guide to Raising and Selling Cut Flowers" by Lynn Byczynski. See Resources for additional books on flower farming and design.

LEARN FROM A FRIEND! Have a friend or relative who's an avid gardener? Trade them weeding time for information. Take time to listen and learn from people who know how to grow.

JOIN YOUR STATE'S FARM BUREAU and/or the local chapter of the National Young Farmers Coalition to attend tours and workshops.

TRY IT OUT! Don't dismiss the amazing opportunity for learning with direct hands-on experience. Doing this will help build your skill level and confidence. Observe and take notes on what works or what didn't work and adjust either your practices or the flowers you grow.

Garden Design Basics

Walking the property and getting to know the layout of our land may have taken a bit of time, but it was invaluable in the grand scheme of things. It gave us the opportunity to observe different growing areas, determine which spots received shade during either early morning or afternoon, learn where the 100% sunny patches stood, identify places which held more water for moisture-loving plants, or the hilly places that provide adequate drainage for growing our stunning lavender plants.

Take a good look at your own growing areas. Look at your location during different times of the day. Where will your seedlings grow best? What does the light look like in the morning? In the afternoon?

Many flowers thrive in full sun, so find that spot. Are there trees around? Generally speaking, a little afternoon shade is always okay. As long as flowers can have at least 6 hours of sun, that's enough to keep them happy and can be considered the equivalent of a "full sun" location.

You'd be surprised to find that a lot of the heat-loving flowers enjoy stepping out of the afternoon sun for a minute. If you have shade, there is a wide variety of flower options out there to plant in the garden — think hellebores, astilbes, hydrangeas, and hostas.

Create Simple Rows

We recommend planting a cutting garden in simple rows to make weeding, feeding, and harvesting tasks easier. It means lower maintenance time for sure, and there's something about straight, neat rows that has a beautiful design aesthetic.

If you decide to plant in rows, many flower gardeners find it successful to design their backyard beds with 3-foot-wide beds along with 2-foot-wide walkways between to create a beautiful aesthetic. Planting flowers of similar heights together in your rows will not only support your design aesthetic, but will also be practical for harvesting your blooms.

If you're interested in integrating cut flowers into existing landscaping, most times the smaller flowers will go in the front, with the taller flowers in the back, so keep this in mind if you're creating circle- or square-shaped beds. Also consider planting taller flowers towards the center, with shorter flowers on the outside.

Planning Plant Quantities

How do you decide how much of everything to grow to create gorgeous bouquets all season long? You have the space to grow, but how do you determine how many plants you'll need to grow in that space? This part can be intimidating, but we've broken things down with some practical tips to build your confidence.

If you're planting rows of flowers, which is our recommendation for cut-flower production, you'll need to measure how long your row is and account for the spacing grid you are using. Our recommendation is to plant almost all flowers at 9 inches apart. Calculating the number of plants you will need for dozens of different flower varieties based on the space you have available can feel overwhelming for beginners and experienced growers alike. We've simplified the process by creating just two separate spacing grids: 6 inch and 9 inch.

We've experimented with plant spacing over the years and have found that 9-inch-by-9-inch spacing is the most versatile approach and one we use for 80% of our flowers. Flowers we plant into a 6-inch grid include: Snapdragons, Statice, Agrostemma, Plumed Celosia, and Ranunculus.

> It's always best to start planning your flower beds by mapping out your growing area.

Think about how you want your flowers to look together, how big you'd like to make your growing beds, where the pathways will connect beds, and other key features. Think about placing your cutting-garden space where it is highly visible to you, so you won't forget to check on it, and also so you don't miss out on its beauty!

Once you've decided on your bed shape, size, and location, sketch the varieties and quantities of flowers you want to grow per row. Graph paper can help you determine how each section connects. We love to plant in a gradient layout for an aesthetically pleasing display that always wows visitors and garden tour guests.

We grow flowers intensively, so we tend to space them more closely than what seed packet instructions might recommend. We do this because the closer you plant flowers, the taller the stems and the more lush they will seem to be. This method is one of the ways we're able to cut hundreds of blooms to make bouquets each week.

You should also plan for filler flowers and greenery for your bouquets throughout the season. As a general guide, equal amounts of focal, filler, and greenery are ideal, but we like to have a slightly larger percentage of greenery. We always need more and never feel like we planted enough, so take our advice and bump up the quantity of annuals, herbs, and perennials that produce beautiful foliage.

Simple tools (pencil, paper, and a calculator) will ensure that you correctly forecast the quantities of flowers to plant, grow, and harvest. Invest planning time before you plant -- and enjoy the payoff when you gather those blooms, like these peak-of-summer zinnias.

Tools + Investments

PSA: You don't need the most-expensive tools to have a successful growing space that produces tons of flowers. There, we said it! Imposter syndrome is something that's talked about a lot in the flower-growing world. Many believe that they need state-of-the-art tools and equipment to call themselves a flower farmer.

We're no strangers to this feeling, either. For over nine years, we used a very simple 10-by-20-foot propagation house to start thousands of seeds and to grow seedlings for our flower farm. Talk about feeling like we're not good enough, as we observed other cut-flower growers, very similar to us in size, using much larger, fancier setups. Now that we're heading into our 10th year, we finally mustered enough extra money to build a new, larger structure, upgrading to a 20-by-49-foot space.

Eventually, we discovered a universal truth: Do what you need to do with the money you have available to invest.

We look for easy-to-use flower-farming tools that are both functional and versatile.
Simple investments into good tools can make all the difference while caring and tending to your flowers.

Supply List

For the backyard gardener, only a few hand tools are needed to manage flower-growing tasks. You don't need all the bells and whistles to be successful. Here are our favorites:

KEY INVESTMENTS TO START

Basic tools: shovels, stirrup hoe, wheel hoe, gloves, buckets, seed trays, seed-starting medium, hand trowel, and floral snips

Flower Seeds

Irrigation or overhead sprinklers

Landscape fabric or weed barrier

Professional seed-starting soil

T5 grow lights and seed racks (You don't need fancy grow lights to start seeds. Fluorescent or LED shop lights from the hardware store work great and are affordable.)

Tiller

LARGER INVESTMENTS TO MAKE
(as you scale)

Backpack Sprayer

Caterpillar Tunnel

Compost Tea Brewer

Coolbot System and Air Conditioner

Drip Irrigation

Earthway Seeder

Greenhouse/High Tunnel

Hoop House

Walk-behind Tiller

Website

Whirlybird Sprinklers

As another example, we still use the original cardboard template we made to burn holes in fabric. There's no need to get worked up if your equipment isn't the best and newest thing on the market. If a farm hack works, use it. Does it work for you? Great!

We've found that there are some completely essential upfront investments and we'll share them throughout this book. Many other investment decisions are either unnecessary, or purchases you should add as you expand your growing space or gain success over time.

Add large investments such as fancy tools or equipment as you see fit, but spread out those purchases and make them in phases. If you are able to purchase everything during your first year, by all means, do so, but if you're just trying your hand at growing flowers, take it slow and steady on the checkbook.

Lower inputs can help the bottom line. It's amazing how much of a positive impact netting a profit in your first year can be for your state of mind. Celebrate small victories along the way, knowing those wins are part of the "secret sauce" that will make you feel successful in your flower-growing journey.

To lay the groundwork for successful cut-flower production, we first till the main annual field and amend it with compost and fertilizer. Then ...

Soil + Site Preparation

The first step is to find out how well your soil will perform for flower growing. County Extension branches often offer free soil testing, so check there first.

ANALYSIS: What do you do if you find your soil lacks certain nutrients? The soil test will identify specific minerals and amendments that your soil needs. You should be able to find most soil supplements at a local garden center or via online sources. [It's best to add any missing nutrients after the step below.]

Once your soil test is complete and you know what to expect from your soil, it's time to use a technique called solarization, which kills off sod or weeds before you till your planting areas. It's easy to solarize! First, water the soil deeply until it's wet. The moisture not only wets your soil, it also helps produce steam, which kills pathogens and weed seeds.

Next, lay a clear, plastic sheeting or tarp over the garden space, burying the edges into the soil to trap the heat. Leave it in place for four to six weeks minimum (up to eight weeks if needed), before pulling it up and tilling your garden bed. The best time to solarize your garden space is during the hottest part of the summer, so plan accordingly.

… we then lay the weed barrier by slightly overlapping rows of landscape fabric and stapling it into the freshly prepared soil before planting seedlings.

Now comes the fun part! It's time to till the solarized patch into garden space. Before you begin tilling, spread any missing nutrients identified in your soil test to the top of the soil. If you're tilling your garden for use the following growing season, we recommend planting a quick-growing cover crop, such as winter radishes, to further enrich the soil.

Weed Suppression

Implement practices or tools to keep the weeds at bay; otherwise, you will be weeding constantly throughout the entire growing season. That would take all of the fun out of growing flowers. It's so defeating when the weeds take over, or when you're weeding almost every day and can't keep up. Nobody has time for that! Weeds can take a tremendous physical toll on your body and mind. Don't let that happen!

When we first started out, weeds in the garden nearly ended our marriage. We weeded nearly every day, and barely held onto the flowers we were growing in our little 10-by-20-foot space. We attempted to suppress weeds by top-dressing with wood mulch, but that only ended up making matters worse, because it introduced a new weed pest: bindweed.

Adam disc-tills the farm's fertile, loamy soil under an idyllic blue sky in preparation for planting one of our many wildflower installations.

We were at a breaking point and knew we had to do something. Thankfully, we love research and we spent a lot of time trying to figure out what we could do. Once we saw other farmers growing in landscape fabric, we knew immediately that this was the way to reduce the amount of time and insanity we spent weeding our poor flower garden.

A frequent question we're asked is: How long does your landscape fabric last? Many of the cheaper, thinner options found at hardware stores just don't do the job and usually only last one season. The tougher, more durable fabric we use generally lasts us about seven years, which we think is very reasonable.

Eventually, it does end up getting a bit brittle and will tear, which is when we replace it. Other great, cost-effective options for weed suppression include pine straw, wood mulch, leaf mulch, common straw, grass clippings, and cardboard. Straw mulch is one of our favorite weed-suppression methods for use on dahlias and also for plants in our raised beds. This hard-working weed barrier adds a lot of organic matter back into the soil, while also helping with moisture retention, which is critical in drier climates. It's great for holding moisture in raised beds, because they need more frequent watering.

A word of caution about wood chips used as mulch: Make sure the chips have begun to decompose before using the material. They are fine as top-dressing, but if tilled into the soil wood chips will tie up nitrogen at the root level, causing stunted growth.

Weed Management

It's much easier to hand-remove weeds peeking out of a tiny little hole in the landscape fabric than toiling over a large space needing weed removal. Even though the use of our landscaping fabric reduces hand-weeding tasks, it's important to keep a regular weeding routine, because it's easy for weeds to get out of control throughout the season.

Our normal weeding cycles occur at two-week intervals once seedlings are planted. Even though you may have a weed barrier in place, do not miss the window to weed.

There are other ways to manage weeds with great tools such as a wheel hoe, stirrup hoe, and flame weeder. The wheel hoe is an invaluable tool we use mostly for our mass plantings of sunflowers, and other fast-growing, open-field flowers. The flame weeder should be used on your garden before planting. Not only is it fun to say you use a flame weeder, but it's very effective to use this to prevent weeds up front, especially in a high tunnel or greenhouse. Use with caution!

Fertilization + Compost

Fertilization and compost aren't typically the most fun discussion points when talking about growing flowers, but soil health and adding nutrients back to the soil shouldn't be a step that is missed. Over time, your soil will gradually lose its nutrient-rich content, as plants pull a lot out of the soil during their natural growth cycle.

We initially use a basic 10-10-10 fertilizer to give our beds a little boost and help our flowers bloom. We spread it out on top of a tilled bed, and then re-till it back into the soil to distribute the fertilizer better in the soil. We do this before the fabric is secured in place and planting begins.

We also include composted manure into our soil. It helps that we have cows on the property to provide it! However, you don't have to have livestock to achieve the same goal. You can buy composted manure and amend it into the soil through tilling or a broad fork. Other sources of compost include your household compost bin and the leaves, wood chips, and animal manure that build healthy compost. Be sure to educate yourself about composted versus "hot" manures before you add anything to the compost bin.

Cover crops are a really great way to add nutrients to your soil, improve soil tilth, and to provide a natural weed barrier. After gardens start to fade in late summer/early fall, we till the soil and plant a cover crop to grow in early fall through winter. We are big fans of

It's gratifying to see this aerial view of our production rows because it illustrates the benefits of planting compactly on a 9-inch spacing grid, leaving little room for weeds to grow.

winter radishes as a cover crop. As the radish grows, its roots expand into the soil. When it dies off, all the organic matter has penetrated deep into the soil. But beware; radishes stink to high heaven when they're rotting. If you are working in your garden and find yourself asking, "What smells like trash?!" It's the radishes!

We also really love buckwheat as a cover crop, because of its quick turnaround time, typically blooming out from seed 35-40 days after planting. It's a great weed suppressor, thrives in poor soil conditions, improves soil tilth, and most important for a working flower farm: we're able to use the flower in bouquets as a sparkly little detail. The flower has a soft, white color, and a lovely fragrance as well.

Vetch, red clover, oats, and wheat are also great cover crops that we will occasionally incorporate into our soil management practices at the farm. Remember, healthy soil is the foundation of a great growing space, so consider these options for managing your soil.

Some annual flowerbeds are cut until frost. In those cases, there isn't enough time to plant a cover crop, so we let the annual flower crop die in place and wait until spring to clean up. That way the soil is held in place and not disturbed before winter dieback. In the spring, as soon as the ground can be worked, the beds are cleaned and soil is lightly tilled, and we plant a quick-growing cover crop like radishes.

You can implement cover cropping into any space size — acres of space aren't necessary in order to accomplish this. In fact, most of our cover cropping is concentrated in smaller growing areas.

Raised Beds

An approach for those with limited space, or those just beginning to grow flowers, is the creation of raised beds. They can create a beautiful and functional growing area right outside your door.

Raised beds are also perfect for new gardens, especially if your soil is not ideal for growing flowers. They provide a great way to start growing flowers, while allowing extra time for improving the soil in other potential growing areas.

From a maintenance standpoint, these beds are especially nice. They're a bit easier to weed and tend to than maintaining open-field plantings. Watering is also generally easier, because raised beds are typically closer to a water source. One thing to note is that raised beds need to be watered more than flowers planted in the ground.

Because we intend to use our raised beds as a space for dahlia breeding, they've helped us to keep an extra watchful eye as we document observations, note breeding varieties, and manage the rest of the process more effectively. They're also just gorgeous! The addition of arches to support vines, and maybe even some bistro lights surrounding the space, make for a beautiful effect.

MATERIALS NEEDED TO BUILD RAISED BEDS

For eight 4-by-8-foot raised beds in a 30-by-40-foot area

For each raised bed, you will need the following:
(3) 2-x-12-x-8 treated lumber
(4) 4-x-4-x-4-inch treated lumber
(4) 6-x-6-inch steel corner braces

ADDITIONAL SUPPLIES:
200 feet of 6-foot or wider landscape fabric to lay at the base for weed barrier
(100) Landscape pins to secure the fabric in place
(160) 3-inch exterior screws
(48) 2-cubic-foot bags of wood chips
(160) 1-cubic-foot bags of topsoil
500 pounds of chipped limestone road stone
Either drip irrigation for each bed, or an overhead watering apparatus
Optional decorative arches for growing vines
Seeds of choice

STEP ONE

DRAW out the raised bed plan on paper. Allow for at least 3-foot-wide paths between each bed.

STAKE each of your four corners with wood. Tie string between stakes to frame out your area and to help identify level.

INSTALL landscape fabric as a weed barrier.

STEP TWO

SPREAD gravel over the entire area with a landscaping rake.

PREPARE the lumber. Each 8-foot board will form the sides.

CUT an 8-foot board in half to form two 4-foot ends.

STEP THREE

With your drill and screws, begin securing lumber at the corners, using pieces of the 4-x-4-x-4-inch lumber as a brace.

STEP FOUR

Lay six bags of wood chips in the bottom 1/2 of each raised bed, using the landscaping rake to smooth out.

Finish by adding 20 bags of topsoil to fill to the top. The soil will settle over time, so you may want to add more before planting.

RAISED BEDS

CHAPTER TWO

Choosing the Right Flowers

THERE ARE SO MANY GORGEOUS FLOWERS THAT ARE PERFECT FOR CUTTING and gathering into bouquets. It's so hard to narrow down a list of the best ones, because when you first start dreaming of your own flower garden, you want to grow everything, especially all the most beautiful flowers.

Before we even first purchased our farm, we started dreaming about our future flower garden and what we would grow. As we pored over seed catalogs, we made lists of the prettiest varieties.

Of course, we gravitated towards romantic-looking blooms and unique collections that could bring added interest to a bouquet, as well as other unusual choices you wouldn't find in online or grocery store bouquets.

Any guesses about one of the first flowers we picked to try growing from seed? Lisianthus. That's right. If you aren't in the know about this flower, the low-down is that they're super-duper pretty, soft and romantic, but they're definitely not an easy or practical flower to grow from seeds when first starting out.

We quickly learned that lisianthus is hard to start from seed, and they take several weeks to germinate. Once they've germinated, they're pretty finicky to care for before you can transplant them in the garden. Then, they take weeks to mature into a nice-sized seedling. In the beginning, we tried to start them from seed. After two seasons, we took a break from our seed-starting efforts and purchased lisianthus as plugs, which are small, established plants.

We're sharing this story, because we want everyone — especially beginners — to have success with seed starting. It's good to experience early success to gain the confidence you'll need to keep progressing as a cut-flower gardener or grower.

The Ten Easiest Cut Flowers

When first starting out, it's great to select flower varieties that are relatively easy to grow from seed. Choosing these sure winners for your cutting garden sets you up for positive results from the beginning. You'll gain confidence with each prolific and productive bloom, and you'll enjoy the beautiful rewards as you master the art of growing flowers.

For a quick reference, we've compiled a list of the easiest flowers to both start from seed and grow in the garden with great success. These are flowers that we love growing each and every year, because they're almost foolproof to germinate and grow. Even as we enter our eleventh year of growing, we keep these easy-to-grow annuals in our lineup of classic PepperHarrow flowers.

These varieties can typically grow in any zone or climate, but be aware of your environment, keeping in mind rainfall, temperature, as well as cultural conditions, such as sun or shade exposure. Many of the flowers on this list do best in warmer temperatures in full sun (at least six hours of sunlight).

Because of the ease of germination, starting any of these flowers from seed is not a difficult task and can give beginners a comfort level with growing while building confidence. From there it's easier to branch out and try some of the more complicated flowers to start from seed.

> Growing these beautiful cut flowers is a snap. We find that they require minimal maintenance or feeding, aside from needing a bit of water after they're planted.

Also, from what we've observed, they tend to withstand a bit more intense heat and stress than most of our other cut flowers.

We also find these varieties to be some of the most profitable cut flowers we grow, because they keep blooming and producing all season long. No matter how hard we cut them in order to produce our weekly bouquets, it just makes them bloom more vigorously. Their resilient nature and ability to continually produce blooms all season long make them invaluable on a production flower farm.

ZINNIA

Zinnia elegans | Grows 3-4 feet | Needs full sun | Our Favorite Variety: Benary's Giant

WHAT WE LOVE: Their assortment of crayon colors and replenishing abundance. Zinnias are one of our most favorite flowers on the farm and have become a summer staple for bouquets.

STRAWFLOWER

Xerochrysum bracteatum | Grows 2-3 feet | Needs full sun | Our Favorite Variety: Fireball

WHAT WE LOVE: Strawflower is a fun filler flower that's loved by farmers' market customers. From white, blush, and apricot to the red options, there's a nice selection to grace any bouquet.

SUNFLOWERS

Helianthus | Grows 5-6 feet | Needs full sun | Our Favorite Variety: Pro-cut Gold

WHAT WE LOVE: Sunflowers are available in a range of colors and bloom sizes. We grow green-centered sunflowers in late spring, transitioning to the dark-center varieties for summer and early fall, ensuring there's a perfect sunflower for our bouquets all season long.

Cosmos bipinnatus | Grows 4 feet | Needs full sun | Our Favorite Variety: Double Dutch

WHAT WE LOVE: Their delicate, airy, and romantic look. Cosmos come in many colors and are a favorite among our pollinators on the farm.

COSMOS

AMMI

Ammi majus | Grows 36-50 inches | Full Sun/Part Shade | Our Favorite Variety: Green Mist.

WHAT WE LOVE: The way it often softens a look of a bouquet. Ammi is a cold-tolerant flower that can be succession-planted all season long. It's a must-have option for wedding floral design.

Tagetes | Grows 4 feet | Needs full sun
Our Favorite Variety: Jedi

WHAT WE LOVE: The bright, bold pop of orange and their nostalgic, unique fragrance. Marigolds are an ideal addition for fall bouquets and are perfect for drying.

Limonium | Grows 18 inches | Needs full sun
Our Favorite Variety: Midnight Blue

WHAT WE LOVE: A great, versatile flower that can be used in fresh-cut flower bouquets and also be used as a dried flower. Statice holds its bold color once dried.

MARIGOLDS

STATICE

BACHELOR'S BUTTON

COCK'S COMB

Centaurea cyanus | Grows 3 feet | Full sun
Our Favorite Variety: Midnight Blue

WHAT WE LOVE: Their unique, pastel colors and the fact that they're edible. They are cold-tolerant making them perfect for early-mid season bouquets.

Celosia | Grows 28-49 inches
Needs full sun; can tolerate shade
Our Favorite Variety: Sunday Orange

WHAT WE LOVE: Plumed varieties and outrageous, bright colors. Cock's Comb comes in two shapes: plumed and crested.

BLACK-EYED SUSAN

Rudbeckia hirta | Grows 18 inches | Needs full sun | Our Favorite Variety: Goldilocks

WHAT WE LOVE: Their cheery yellow pop of color in summer and fall bouquets. Black-eyed Susans are available in a wide selection of petal forms and bloom sizes.

PERENNIAL

We use snowball viburnum, considered a woody perennial, as a staple greenery and filler flower for early spring bouquets.

ANNUAL

Although a bit on the short side, these field-planted lucky lips snapdragons provide a pop of color in early summer bouquets.

PERENNIAL

Coneflower is valued as a focal flower and for its proclivity to attract pollinators. We love its interesting center, which we sometimes use even without petals in bouquets.

ANNUAL

Larkspur is a multi-use annual, which we include in fresh bouquets as an accent flower. We also dry larkspur for use in dried bunches and wreaths.

Perennials vs. Annuals

Which should you plant? It really depends on your goals. For efficiency purposes, we really like annuals. They're reliable, because most seed packets state the bloom date from when you sow the seed (such as 90 days or 120 days). Annuals bring a whole lot of color into the garden. They pump out the blooms, which is invaluable for bouquet production. We also love them because we know once they're done blooming, it's okay to pull them out, clean everything up and plant something new in their place.

It's beautiful to start a second wave of annual flowers at midseason and enjoy additional productivity from the same planting area where you grew that first wave of flowers. Aesthetically, this plan also keeps our gardens looking gorgeous for visitors to enjoy throughout the season.

PERENNIALS TO TRY	OTHER ANNUALS TO TRY
Columbine	Ageratum
Delphinium	Agrostemma
Dianthus	China Aster
Goldenrod	Gomphrena
Lady's Mantle	Larkspur
Peony	Orlaya
Rudbeckia	Phlox-Cherry Caramel
Salvia	Saponaria
Veronica	Snapdragon

On the other hand, we also have a lot of love for perennials. They provide a reliable bloom source each and every year, especially for valuable flowers we need for bouquet designs. Perennials' predictable bloom time is almost exactly the same from year to year, so we know how to plan what blooms when and can be ready to go with a bouquet design when it happens.

For example, each spring we love using salvia with lady's mantle and ranunculus or anemone as the focal flowers. These flowers always bloom together and the mix of them is a winning combination! Consider how much work you want to invest into growing annuals and perennials. Typically, perennials mark the seasons with stunning blooms, but they will have more weeding requirements, since they're permanent in the landscape. Annuals give you lots of bang for the buck, but they are time-consuming, since you have to grow them from seed and replant them each year. After a few seasons, we're confident that you'll discover the best annuals-to-perennials ratio for your garden or flower farm. Our crop mix consists of nearly 80% annuals and about 20% perennials.

FOLIAGE   FRAGRANCE

Various types of herbs are staples for use in bouquets all season long.

Heirloom roses such as Lady of Shalott and Strawberry Hill deliver intoxicating fragrances.

Flowers by Use

How each flower will be used is an important quality we consider during selection and planning. Does the flower or greenery have additional uses that will create more perceived value? For example, fragrance is a bonus in blooms, as well as in greenery.

FOLIAGE: Foliage and greenery act as an anchor to your bouquet and give a fuller look. Sometimes it's hard to find enough foliage to include, depending on the season. We look to herbs as a solution, including rosemary and apple mint, but we also use cultivated varieties of foliage such as Persian cress and atriplex.

FRAGRANCE: Infusing fragrance through the use of scented flowers or herbs is a great way to enhance a bouquet. Whenever possible, we try to include either fragrant flowers or scented herbs to create something special and unique for our customers.

EDIBLE: Adding edible flowers and herbs to your growing lineup is a great idea, because there are so many uses in the kitchen for beverages, culinary dishes, and baked goods. One of our favorite inclusions is 'Mojito' mint for our summer bouquets. We encourage customers to use it in their drinks later in the evening.

DRYING/PRESSING: We look for flowers that dry well for bouquets and/or wreath-making, and that can be pressed for floral art. As a part of our season extension on the farm, these are important uses that create added value during the fall and winter months. Our favorite flowers to dry are marigolds and celosia, because of their beautiful and vibrant colors.

EDIBLE

We grow edibles like lavender for use in tea blends and baked goods, and to sell to local chefs.

DRYING/PRESSING

A rainbow of dried flowers can be used for bouquets, crafts, or accents to a holiday wreath.

TRY GROWING THESE ANNUALS WITH SPECIAL ATTRIBUTES

We appreciate these flowers for their sensory and design-specific traits

FOLIAGE	FRAGRANCE	EDIBLE	DRYING/PRESSING
Dusty Miller	Basil	Basil	Ammobium
Ferns	Dame's Rocket	Borage	Baby's Breath
Herbs	Eucalyptus	Calendula	Calendula
Hosta	Honeysuckle	Chrysanthemum	Celosia
Lilac	Hyacinth	Dandelion	Craspedia
Ninebark	Lilac	Lavender	Delphinium
Peony	Mock Orange	Pansy	Lady's Mantle
Thornless Raspberry	Narcissus	Perennial Phlox	Larkspur
Tree branches	Phlox	Rose	Lunaria
Viburnum	Rosemary	Sunflower Petals	Penny Cress
Willow	Roses	Sweet William	Saponaria
	Scented Geranium	Violet	Statice
	Sweet Annie		Strawflower
	Sweet Pea		Sweet Annie
	Sweet William		
	Tuberose		
	Tulips (Doubles)		

Design-Driven Flower Farming

Buyer psychology is a very interesting science and is something to which you should pay attention. Observe what you see happening in the market via social media, what your customers gravitate towards, and keep experimenting with new varieties.

During the past decade of growing and selling flowers at farmers' markets, we've noticed one thing in particular: Customers buy flower colors based on the season. We have observed our customers tended to buy pastel colors in the spring, bright colors in the summer, and warmer, autumnal colors in the fall. Now, our seed-starting schedule is based on this consumer behavior.

Our spring color palette typically revolves around more pastel shades, such as pink, lavender, and peach; however, our customers are drawn to orange and pink together in a bouquet, so this is one exception. Think pink ranunculus paired with orange tulips in a vase. So, so lovely!

During the beginning and height of the summer months, we grow as many sunflowers as possible.

To the bright yellow impact of sunflowers, we add vivid pops of purple, hot pink, dazzling orange, and all the rest of the saturated rainbow colors. Primary hues and bright, bold colors are all the rage this time of year.

In the fall, the petal spectrum warms up and customers are drawn to richer jewel tones such as dark purple, dark pink, burgundy, red, and burnt orange, accented by organic-looking taupe and neutrals. All of these colors echo autumn trees as their leaves turn colors.

From mid-summer to fall, we have a prolific, standout flower: Dahlias. With more than 12,000 plants, our dahlias burst into bloom with so much vibrancy and for so long that they justify their own section! In the seasonality of color, dahlias truly stand on their own. Our dahlias first emerge in early July and continue blooming until mid-October. From soft, romantic blushes and bright fuschias to the autumnal colors of fall, we grow a rainbow of dahlias in many different shapes and sizes.

As we reach the end of our growing season and look towards winter, we naturally start to gravitate towards using all of the varieties we have harvested at the height of the season and prepared for drying. We also shift our focus to using lots of fresh-cut evergreens for wreaths and centerpieces for the holiday season.

SPRING

SUMMER

AUTUMN

WINTER

SPRING

Apricot parrot tulips dazzle with their unique painterly and multicolored petals.

With a distinctive purple-black center, the anemone is a popular early spring flower.

Long-lasting, pastel-colored ranunculus are a must have for spring bouquets.

Fluffy and romantic, peonies are a classic and picturesque early-to-late spring flower.

Double daffodils with their ruffles and delicate fragrance bring charm to early-spring designs.

An ode to spring includes pastel mixes of garden roses and anemones.

An unforgettable early-spring flower, hellebores are available as frilly doubles or single forms.

Mixes of allium, columbine, shepherd's purse, snowball viburnum, tulips, and stock.

SPRING

A summery mix of black-eyed Susans, zinnias, ageratum, strawflower, and ammi.

Zinnias are a warm-season staple producing a mix of vibrant colors and forms.

Interesting star-shaped astrantia blooms in early summer and is also a good dried flower.

Yellow zinnias and sunflowers are paired with snapdragons, statice, strawflower and baptisia.

SUMMER

Frilly, double lisianthus blooms arranged with the bright pops of color from seasonal ageratum.

Orange and yellow Benary's Giant zinnias make their way into many summer bouquets.

Bountiful harvests of lavender are gathered from mid-June through July.

Papery strawflower add interest to bouquets, in many colors, including blush and burgundy.

An early dahlia bloom, Linda's Baby, looks pretty against Jenn's velvet dress before an event.

DAHLIAS

Cafe Au Lait Rose, Beatrice, Sweet Love, Cornel, Peaches 'N Cream and Ivanetti dahlias.

Peaches 'N Cream, Cornel, Mister Frans, Coralie, Ivanetti, and Boom Boom White dahlias.

DAHLIAS

A romantic color combination features Sweet Love and Boom Boom White dahlias.

Cafe Au Lait Rose, Cornel, Ivanetti, Jowey Mirella, Hollyhill Black Beauty, Linda's Baby, Brown Sugar, Jowey Winnie, Genova, Miracle Princess, and Sweet Love dahlias.

An autumnal mix of zinnias, black-eyed Susans, lisianthus, combed celosia, and spiked celosia.

Jedi marigolds are always something to look forward to during the bounty of fall.

Most of our fall bouquets contain a mainstay flower, such as combed peach celosia.

Bright yellow zinnias with fuchsia tones of celosia, and Arabian Not dahlias.

AUTUMN

AUTUMN

Porcelain doll pumpkins grace the farm and become a seasonal addition for on-farm sales.

Fall dried bunches of celosia, marigolds, strawflower, and ageratum.

A collection of ammi, globe amaranth, salvia, zinnia, marigold, and Peaches 'N Cream dahlias.

Forcing amaranth bulbs in preparation for the holiday season is a great fall activity.

Early winter brings preparation for the holiday season ahead, as our days are filled making festive, seasonal and sometimes non-traditional fresh evergreen wreaths. This is our period to be creative, and to relax and enjoy our downtime.

WINTER

WINTER

Fresh Fraser and Douglas fir, pinecones, incense cedar, Oregonia greenery tips, magnolia leaf accents, and dried oranges.

Douglas fir, magnolia leaves, juniper, winterberry, and pinecones.

Douglas fir base with dried flower accents of burgundy celosia, winterberry, strawflower, globe amaranth, salvia, pistachio seed pods, lemon cypress, oranges, and Chinese lanterns.

Douglas and Fraser fir base with red cedar, incense cedar, Oregonian greenery tips, winterberry, and twigs.

CHAPTER THREE

Growing the Flowers

Based on more than a decade of experience, we've landed on a multitude of methods to grow acres and acres of beautiful flowers through the season. There are many different ways to grow beautiful flowers, so as you read through this chapter, keep in mind that you will gain your own growing expertise and find what's best for you. Be flexible, learn everything you can, and adjust accordingly.

Can you already visualize the beautiful bounty of flowers you're about to grow? This is where the work begins as you flex those flower-growing skills and put them to the test. We'll share all of our best methods for not only growing, but maintaining gorgeous blooms.

Starting from Seed

Seed starting may feel a bit overwhelming or just too complicated to undertake. Some of our students have felt as if they couldn't be successful unless they invested in all the bells and whistles (aka tools and supplies). The good news: You don't have to follow every process to a "T," nor do you have to invest in the fanciest tools to be a champion seed starter. It's okay to experiment and deviate a little from what you read on the back of a seed packet, and still enjoy success with growing seeds.

If you're just starting out, the best plan is to start seeds that require very little TLC. Select easy-to-grow blooms that ensure "beginner's success" while you build confidence. Try zinnias, bachelor's button, celosia, cosmos, and amaranth. These beauties grow so easily, you can practically throw the seeds in the ground and watch them take root, rather than starting in seed trays.

Nearly all of our cultivated cut-flower varieties at PepperHarrow are started in seed-starting trays. Our most reliable method is to use 72-cell seed trays. The large cell trays work better for seed starting, because they allow for a bit more moisture to hold in the planting medium, which helps seeds to germinate with higher viability.

 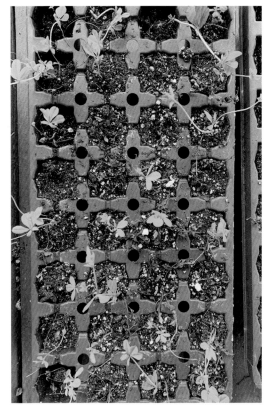

Transplating, or 'bumping up' seedlings into a larger cell is sometimes necessary.
Left: Money plant seedlings are ready for their new space in a 50-cell tray.
Right: Early sweet pea seeds planted in deep-cell trays.

We use Pro Line Mix from Jolly Gardener as our preferred seed-starting medium. It's a professional-grade mix of Canadian sphagnum peat, medium perlite, and vermiculite, and it has a fine texture. You can find seed-starting mixes in smaller quantities at nurseries and garden centers in early spring.

We start our flower seeds in our basement, in a space we have come to call our "Middle Room." We use wire-rack shelving with grow lights. This layout produces hundreds of seedlings — enough to plant 1.5 acres of cut flowers.

For lighting, we use professional-grade grow lights, but we've found that you don't need to be that fancy. Something as simple as T5 shop lights are good enough to get your seedlings started. Seeds just need some type of light source to germinate.

In general, leave grow lights on for at least 12 hours each day; the corresponding 12 hours of darkness gives the seedlings a break. It's not too difficult to manage if you purchase a simple timer and set lights to turn on/off at 12-hour intervals. This process stimulates plant growth because it emulates their growing pattern outdoors.

Another important detail requiring attention is the temperature in the growing area.

It's important that the temperature remains fairly consistent to ensure germination. Extreme temperature fluctuations may impair the germination rate on your seedlings. Varieties that can take a bit more cold will germinate in cooler temperatures, so it's natural to start those first. Our growing space starts out cooler during late winter and warms up as the season progresses outside. As always, check the seed packet, or look up best germination temperatures for the seeds you are starting. This is a lot of homework if you've never started seeds before, but it gets easier as you go. Keep notes of the seeds you start, recording germination temperature, days to emerge, as well as days to bloom.

Direct Sowing in the Soil

Success with direct sowing really depends on the quality of your soil. We would love to do more direct sowing in the field, but the ground we're working on used to be horse pasture and was long neglected, so even after 10 years, we're still battling old weed seeds. That's tough competition for direct-sown annuals. Most of the cultivated varieties of flowers we grow are started in seed trays, which are then transplanted outside once established. There are a few seeds that establish themselves, grow quickly, and beat out their weed competition, which makes them perfect for direct seeding.

Flowers we occasionally direct-sow are zinnias, sunflower, cosmos, and larkspur. Using bulk seed for direct-sowings is also cost-effective. When planted en masse, these flower installations can create a stunning, abundant backdrop for photographers who want to capture client portraits in the flower field.

We also sneak zinnia seeds into a few empty spots that inevitably appear, such as the random holes that become available in our landscape fabric when other seedlings fail. It's always best to have a flowering plant growing in that garden spot rather than weeds, so be prepared to direct-sow annual flower seeds in those locations.

COSMOS

Cosmos are among our most-prolific bloomers on the farm. We love this carefree, cheerful annual because it's so easy to start. Cosmos are absolutely stunning in the garden with their delicate blooms that sway in the breeze. For specialty varieties, we sow one seed per cell in a 72-cell tray and water the tray, never letting it dry out completely.

SUNFLOWERS

Sunflowers are an easy, wonderful flower to directly sow into the ground. We prefer using a walking seeder with a corn plate to plant sunflower seeds into rows. Once sown, make sure to water regularly and put them on a two-week weeding cycle. You'll be harvesting those gorgeous sunflowers before you know it.

ZINNIAS

The wide selection of colors and varied petal shapes make zinnias another favorite for direct seeding. Although they don't germinate as quickly as cosmos or sunflowers, they still do a great job of outpacing small weeds that might compete for growing space around them.

Multiple successions of plantings occur throughout the growing season on the farm. Shown here are new rows of amaranth and celosia planted after early-summer zinnias finished blooming.

Succession Planting

Have you tried to grow flowers in the past, but your garden doesn't seem to yield a steady amount of flowers for a season of making bouquets? Or have you observed that flowers start to get tired around August or September, slowing down their productivity? It's possible to correct both of these problems with a practice known as succession planting. The overall concept is simple: You start seeds multiple times throughout the season to enjoy continual blooms.

During planning and preparation for our growing season, we first separate seeds by those that are tolerant to cold weather versus seeds that have a faster turn and prefer

1

JANUARY
Sow lisianthus for May planting.
Pre-sprout ranunculus and anemone in crates for planting in February.

FEBRUARY
Sow snapdragons, larkspur, bachelor's button, Queen Anne's lace (Ammi majus), agrostemma, stock, and asters for March planting.

2

MARCH
Plant perennials, any specialty seeds, and seeds that love cooler weather temperatures.

APRIL
Start heat-loving, quick-turn (60-day bloom) seeds, such as zinnia, amaranth or marigold.

warmer weather for their germination. We also separate seeds by seasonal flower color as we described earlier.

From there, we plan seed-sowing successions into four sowings by both season and color. Our first sowings are early-spring cool flowers, or slow-growing seedlings such as lisianthus, planted early in March. Our next two successions are bright, summer colors, with the second planting scheduled for early-May, and the third planting scheduled for early-June. The final seed-sowing succession includes varieties that complement the fall color palette, which are sown early in June, and are planted outdoors early in July.

We've simplified the process of planning with this four-phase approach. As the result, we've created an average of 90 days from sowing to bloom for both summer and fall. There are always outliers to this strategy, such as slow growers like lisianthus, but 90 days works as a general rule of thumb.

The third and fourth planting successions generally coincide with the hot temperatures of summer. Here are a few succession-planting tips we recommend for this period:

TIP 1. Plant either early in the morning or later in the evening.

TIP 2. Soak the growing area with a sprinkler before planting — or even better, plant after a nice rainfall.

TIP 3. Water seedlings with a sprinkler once again after they're planted.

TIP 4. Water every morning for at least one week to establish seedlings, and keep them refreshed from the heat.

Although popular with many home gardeners and flower farmers alike, we rarely fall-sow hardy annuals, which are flowers that can grow in cooler temperatures; however, it certainly can be done. We've tried this in the past, but found that it wasn't worth the extra labor of managing row covers and doing weed maintenance.

We experimented recently with a mid-July planting of snapdragons in fall colors. While our original intention was to use the snapdragons for fall bouquets, we found they didn't really bloom much. However, the following spring, we discovered that the snapdragons overwintered in place, and we were surprised to have something beautiful to include in our Mother's Day bouquets. It was so nice to have them for early bouquets!

3

MAY

Do a second sowing of summer blooms, including sweet Annie (Artemisia annua), ageratum, gomphrena, zinnia, strawflower, celosia, amaranth, basil, and marigold.

JUNE

Schedule a last sowing for fall blooms such as zinnia, celosia, amaranth, basil, and marigold.

4

JULY

Complete last plantings by the end of July/early-August. In areas where you aren't planting flowers, plant cover crops.

SEPTEMBER

Plant cover crops in areas where summer annuals have been removed.

Flipping Your Growing Space

It's hard for most gardeners to flip their space for a new succession of flowers because it goes against the core values of tending and growing, but this is a necessary step to ensure you have blooms all season long.

Our spring-bloomers are generally the first flowers to be phased out in time for the next succession planting. We start with taking out tulips, anemone, ranunculus, bachelor's button, stock, and snapdragon. We do this by either mowing or weed-whacking spent blooms, removing them to the top of the landscape fabric.

Once this step is completed, we pull the fabric up and retill a row or entire space — depending on the replanting area — adding a bit of granular fertilizer back to the soil. Then we lay the fabric back down on the space, pin it in with staples, and replant with the next succession of seedlings.

Transitioning the Seedlings Outside

Transitioning seedlings from the cozy, seed-starting, indoor environment to the occasionally harsh, outdoor elements should be done carefully, because after being in such a controlled environment, those young plants are more sensitive to the sunlight and strong winds. This entire process is one that gardeners call "hardening off."

If you have just a few plant trays, we recommend bringing them outside to a protected spot for at least one hour the first day, returning them back inside after that initial hour. Gradually increase the time you do this each subsequent day.

If you're growing on a large scale and have large quantities of seedlings, we recommend transitioning them outside to a shaded area, protected from the wind. We place trays of seedlings in an unheated hoop house on mesh tables, and we cover the top of the hoop house with shade cloth. Small seedlings are left in this environment for about a week, and then they are transitioned outside into full sun. Keep them watered well — at least twice a day (morning and evening), or three times daily on very warm days, with an additional watering around noon.

How do you decide what to plant outside and when? We make a game plan, beginning with the seed starts that can take a bit of cold temperatures. We have found that it's okay to plant seedlings that are tolerant of colder temperatures between 60-75 days before your zone's last frost date.

We start with ranunculus and anemone, quickly followed by snapdragon, stock, bachelor's button, ammi, corn cockle, and other flowers that can take a bit of cold. We wait to plant everything else around the time of our last frost date (sometimes we cheat a little and plant outdoors up to two weeks before the last frost). It's a gamble

Hardening off seedlings before planting them outside helps them successfully acclimate to the sometimes inconsistent temperatures and wind conditions. At right, a liatris seedling is ready to be transplanted into a pollinator garden.

that sometimes pays off, so we encourage you to experiment with some, but not all, of your seedlings. If you plant flowers that are not tolerant to cold temperatures before your area's last frost date, you may lose them if temperatures dip down to 32 degrees F or lower.

We've lost dahlia cuttings at around 39 degrees F, even after the last frost date has passed, so be sure to give them a bit more protection by covering them with a sheet of Agribon row cover. For small growing spaces, we highly recommend an investment in protective covering, which is supported by small wire hoops. This setup protects young seedlings planted out before your last frost date.

To transplant seedlings, we grab a handy-dandy butter knife that's perfect for scooping each little plant out of its cell tray, placing that baby seedling into an empty hole in the landscaping fabric. It's our preference to straddle the middle of the row, working our way forward as we plant, pushing the tray forward as we go over the top of the seedlings we're placing into the ground. This forward motion makes planting go so much quicker. We can usually plant an entire 50-foot row in about 30 minutes. That's six trays of 72 cells, and a total of 432 plants.

Extending the Season

At our small farm, we cherish a little time off from growing in order to reset our minds, and give ourselves a chance to relax, read, create, and strategize for the season ahead. That much-needed break from the constant hustle of tending to our flower fields usually comes after the holidays, into the New Year.

Many growers aim to grow all year long, but our intention is to capture a few extra weeks both early (before last frost) and later (after first frost). This is called season extension or "shoulder" seasons. Season extension is just a fancy term for growing flowers either earlier or/later in the year's cycle in order to extend your growing time.

If space is not an issue, we love the use of hoop houses, cold frames, and small greenhouses to grow seedlings during the shoulder season. These types of structures require more investment and may take up more space, but they are valuable additions once you've established your operation and can justify such investments for future years of growing flowers.

Fabric, Netting and Staking

As we already outlined, we prefer mostly the 9-inch-by-9-inch spacing for planting into the landscape fabric. In order to make the template for this, we took a 3-by-4-foot piece of cardboard, marked holes at every 9 inches, and then burned the holes with a torch. Fireproof tape is added to the edges to protect the opening from burning during future uses.

Once your template is ready, you can begin burning the holes into your fabric. We've found that 4-foot-wide fabric works the best for us. We lay the fabric out on pre-prepped, tilled rows, which match the width of the fabric, then staple them in place. Standing on the side of the row, we place the template at the top, burn the holes, then we pick it up and move down the row, matching the holes to complete the pattern.

In regard to netting or staking, we are not big fans of this method for our flowers. Not only do we think it's unsightly in a beautiful garden, but a lot of the time, the netting just ends up getting in the way. This makes it hard to harvest, and it's difficult to keep tidy until the end of the season.

In some cases, both netting and staking are necessary, but it's easy to go overboard. If you plant in the spacing we suggest, your flowers stay upright on their own. However, the flowers that we do recommend using netting for, especially if you're growing in a windy, exposed area are marigold, lisianthus, and ageratum. The flowers we recommend for staking or the corral technique include delphinium and dahlias. The corral technique involves placing a metal T-post every 8 feet along the outside edges of the bed, and stringing a double layer of baling twine from post to post.

Ways to Water

Watering your flowers should be part of a standard routine; however, you really have to pay attention to how much moisture, or lack thereof, your plantings receive in any given week. We keep track of rain totals, and routinely check the soil to see how much moisture is being retained. If it's super dry, we give our plants a good soaking. If they're super wet, we pray for some wind and warmer temps to dry the soil.

When we first started out, we used a simple overhead sprinkler to irrigate our growing space. This is a great option for smaller areas, and is very economical.

The only downside is that overhead watering like this can make some of your flowers more susceptible to diseases like powdery mildew and black spot.

After our initial two growing seasons, we progressed to using drip irrigation. Drip irrigation is a series of tubes, called drip line (or tape), which you hook up to your water source. The drip tubes have emitters on them that allow water to flow out and onto each plant to keep them watered. Drip irrigation is placed under the landscape fabric we grow in, and has a master line that feeds out to the flower fields from the main water outlet.

Drip irrigation became problematic for us for a few reasons. The first thing we struggled with was maintaining enough pressure in our lines to supply water to our many, large growing spaces. We added booster pumps to help push additional water pressures out to the flower fields, but this ultimately ended up not working and created a maze of complicated-looking devices. It was a total mess! We also found that the drip tape would water inconsistently under the fabric. It would kink very easily, and finding the location underneath fabric was a chore within itself.

The other issue was the cleanup, because rolling up what felt like miles of drip line back onto the spool each season was a dreaded task. We were holding our flower farm together with very few employees, which is a pretty common situation for young farms. Full candor here: some seasons, the irrigation system didn't get put away, and we would end up with those random drip irrigation lines flapping in the winter winds, which was embarrassing and unsightly.

After our struggle with drip irrigation, we moved to a simpler method of watering through the use of overhead wobbler sprinklers. The wobbler sprinklers are incredibly efficient, and are designed to distribute the water from the head in a raindrop-like fashion — rather than the spray of overhead sprinklers. Standard sprinklers waste a tremendous amount of water through evaporation with the mist you see flying away with the slightest breeze.

The design of the overhead wobbler head is pure brilliance with its water-saving efficiency. Our farm is now irrigated with all wobblers and, as an added bonus, the system is pennies on the dollar compared to the cost of a drip-irrigation setup. We've also found that our water usage has not increased with the switch from drip irrigation to wobbler irrigation. Our technique for overhead watering effectively is to water after harvest, alleviating any concern of cutting blooms when wet.

Feeding the Flowers

Compost tea and any other foliar feeds provide excellent nutrition and a boost for your blooms. We recommend spraying compost tea every couple of weeks.

We also occasionally give our blooms a boost with fish emulsion or sea kelp. Mostly, we use these for our most-expensive crops, such as ranunculus, anemone, and dahlias.

Powdery mildew is a persistent, pesky disease. From what we've observed, via a side-by-side comparison, a direct application of compost tea every two weeks dramatically reduces the occurrence of powdery mildew in our zinnia crops compared to zinnias that don't receive it. Compost tea is also known to help plants resist bug pressure from pests.

Pest Management

We have many pests here on the farm, so a good integrated pest-management program is important for control. Bad bugs can be the bane of your existence during the growing season. The pest battle may vary from year to year, but make no mistake: We always have bug pressure. It's important to have a plan for how to deal with them, because they can destroy your beautiful flowers in no time.

What kinds of pests do we battle? The most intense bug pressure for our flowers comes from Japanese beetles, cucumber beetles, tarnished plant bug, and thrips. Eeek! Even putting their names on this page creates anxiety. We wage war with each of these pests in one way or another each growing season. (Learn more about specific products in our Resources section.)

Our pest management plan is as follows:

> **RELEASE NEMATODES IN TWO-WEEK INTERVALS** starting in April when soil is workable, but hasn't yet warmed up. We usually spray three to four applications at the start of the season.

Knowing the difference between a good bug or bad bug is critical to pest management. Left: Japanese beetles are one of the invasive species of bad bugs that can decimate flower crops. Right: Monarch caterpillars are often found on milkweed plants and are good bugs that should be protected.

SPRAY NEEM OIL TWICE EARLY IN THE SEASON, (two weeks between applications) on roses and shrubs for early pest prevention. Spray dahlias three to four times (two-week intervals) early in their season for prevention. Neem is an insect growth regulator and does not work with a single application. Subsequent applications are required for effective control.

PYGANIC IS USED TO MANAGE TARNISHED PLANT BUGS, Japanese beetles, cucumber beetles or squash bugs. Pyganic has great immediate "knock down" power. Spray in two-week intervals at the first sign of detection on sticky trap paper.

SPINOSAD IS USED FOR THRIPS, CATERPILLARS, AND WORMS. Spray in two-week intervals at the first sign of detection on sticky trap paper.

NEEM OIL or insecticidal soap helps with aphid control.

CHAPTER FOUR

Tie a Ribbon

GROWING FLOWERS IS VERY REWARDING AND THERAPEUTIC, BUT THE anticipation of cutting those blooms to bring inside to arrange and enjoy, share with friends, or bundle for market are equally gratifying reasons to plant a cutting garden.

Throughout the growing season, usually on a whim, we venture out into our beautiful garden where a bounty of blooms greets us with their potential as design elements. Taking a small bucket of water and a pair of floral snips, we select flowers that catch our eye. As we gather them, we're sure to include at least three different types of greenery, a few special attention-grabbing blooms that add extra sparkle, several accent flowers, and usually some sort of grass or other airy-wispy goodness.

Cutting from our garden and arranging flowers is the best and most meaningful way we spend time together. We put on a little music, talk, relax, and enjoy each other's company — not to mention the stunning vistas across our magical property — all while having fun. It's the perfect date night, designing with the gorgeous flowers we grew. Somehow, we forget the investments of time, money and hard work that led to these blooms, and it all feels worth it.

Cutting Your Blooms

The sight of a garden exploding in a rainbow of color and prolific blooms is simply the best, especially when you know you're going to cut lots of flowers to fill several buckets or vases, all with the goal of creating a beautiful arrangement or bouquet to share. There's such a feeling of awe in knowing our hands planted the seeds and tended to the seedlings, as we cut those stems to enjoy.

Before you cut your first flower from the garden, check to make sure your floral snips are clean and ready for use. Dirty floral snips can promote disease in cut flowers, so make sure you take a moment and clean them with soap and water.

Plan to cut your flowers during the coolest part of the day (usually during the early morning), and bring a tall container of tepid water along with you. Place your stems into the water immediately to help them hydrate.

As you cut blooms off of your plants, make sure to have a nice, long stem (ideally at least 24 inches long). For many annuals and perennials, cutting actually stimulates a plant's ongoing growth and flower production. Just keep in mind that plants feed off of their foliage, so when harvesting, don't cut a stem completely to the ground. Leave at least one-third of the plant at any given time.

GOLDEN RULE FOR GARDENERS: the more you cut flowers, the more you get! Be sure to deadhead blooms before they start producing seeds. Like cutting flowers, the act of deadheading also promotes new growth and flowering.

We typically give our blooms a little rest period after cutting them. This doesn't always work out, especially if we're pressed for time; however, flowers love it most when you give them at least four to six hours to rest inside in a cool place out of direct sunlight.

Cutting early in the morning, using a bucket filled with fresh, room-temperature water, Adam harvests thousands of stems of celosia from the flower field. A favorite accent in bouquets, celosia is one of our leading crops. We also use it as a dried flower in bouquets and wreaths.

Give your zinnias, strawflowers, and globe amaranth plants the wiggle test before harvesting stems. Grab the stem and give it a little shake. If the stem is NOT 100% stiff, do NOT cut the flower. These flowers aren't ready for cutting, and will flop over and wilt within hours. Not fun!

After your flowers have had a short rest period and you're ready to start arranging, make a fresh cut to the bottom of the stem and add a floral preservative to the water to further prolong their vase life.

Hydration

This should be the flower's first drink, especially when harvest occurs during warm temperatures. While placing just-picked flowers directly into water is the standard rule, there are some additives that can also help with his process. Hydration options include:

OVB: Use in water when harvesting field-grown crops during higher temperatures. Flowers can be left in the solution for three to four days.

PROFESSIONAL #1: Great for roses, hydrangeas, and woody ornamental plants. Flowers can stay in this solution up to three to four days before transferring into holding solution.

Not all varieties can handle processing solution. See the Postharvest Care book by the ASCFG in our Resources.

Holding Solution

You'll want to store all of your flowers in a holding solution to prolong vase life, keep blooms from blowing open, and to help maintain clean water in the bucket.

CVBn (slow release chlorine tablets): A must for hairy-stemmed flowers such as rudbeckia, sunflowers, zinnias and marigolds; it's also good for dahlias. Most flowers can be cut directly into this solution. It should stay active for one to three days, depending on temperature.

BULB T-BAG: For blooms from corms, rhizomes, tubers, and bulbs. Use as first drink when you process the flowers and until they are sold. Prevents premature yellowing and bud stagnation, and improves vase life. For dahlias, we combine one CVBn tablet with a Bulb T-bag per bucket when we harvest.

Garden-Inspired Design

Playing and creating with the flowers we grow is one of our most-favorite things about having a flower farm. Over time, we've experimented with lots of color, texture, and flower combinations in our bouquets, and we have a pretty good idea about what looks pleasing to the eye.

When we first started supplying flowers for weddings and events, it took us a while to get used to people calling them "wildflowers," because almost all of the flowers we use for our bouquets are cultivated and grown from seed by us, not something we pick from nature, or have grown wild.

After a while, we realized our customers just meant that our designs have more of a natural aesthetic than what they would find elsewhere. Such a wonderful compliment! We learned to embrace our more natural floral design style wholeheartedly.

Because we use a diverse array of botanicals from the garden and farm for our designs, which expresses a natural design aesthetic, we often call our floral style garden-inspired. From poppy pods and foraged branches to other interesting, not-so-common flowers, these elements provide tons of interest for our floral designs. For bouquet greenery, we love using herbs from the garden and branches from our woody perennials, such as snowball viburnum, and ninebark.

> The idea of bringing nature from the outside to create organic accents within the home is something we encourage people to do regularly.

Examples of this could be something as simple as cutting a few branches from a tree to add into a vase, or picking a bouquet of herbs to fill jars for the kitchen table. You don't necessarily have to have a large garden of flowers to appreciate and enjoy botanicals. There's so much beauty in simplicity. Another uncomplicated but pretty approach is to use only one flower variety in your design like tulips, daffodils, or cosmos. Actually, any flower looks great when showcased in abundance as their color variations and similar shapes create a singular arrangement.

Your own design style will evolve over time as you practice and play with flowers and other botanical elements. Never forget that what you arrange is an expression of your own inner artist, and is your own creation. It may never look exactly like someone else's style — even if you use another design as inspiration for the basis of what you're trying to achieve. Trust yourself, explore your own style, and have fun creating!

MODERN DUTCH MASTER

Using old Dutch master paintings as inspiration, Jenn selects colorful early-spring blooms to create a dramatic, garden-inspired arrangement. The black backdrop provides a striking, neutral background to highlight this design's exquisite color palette.

A PINK + ORAGNE AFFAIR

Claude Monet was known for his obsession with light in his paintings.
It worked to bring his paintings to life and gave them an ethereal feeling. Using natural light as an accent here, Jenn highlights a favorite PepperHarrow Farm color combination of orange and pink. Featured are ranunculus, Star of Bethlehem, and eucalyptus.

Color Combinations

We often joke that designing bouquets is a flower farmers' eye candy. We love making bouquets for people with sweet color combinations. If done right, the bouquet should literally look like candy.

Our favorite colors to mix and design with, especially for market bouquets, are high-contrast, bold, bright colors such as oranges and pinks with pops of salmon. Another favorite is purple paired with orange and hot pink. Bold colors pack a lot of punch and our customers love them. These color combos have quickly become seasonal bestsellers, both at markets and local retail stores.

On the other end of the spectrum, a few of our customers also appreciate softer color combinations, like peach and yellow; or peach, soft orange and other muted colors; and romantic bouquets with lots of pink and white flowers. Most customers gravitate towards brighter palettes, but we love to offer lots of options to inspire customers to take home fresh flowers every week.

Another design style to consider is monochromatic. With this approach, you would use the same color of a variety of flowers to create your arrangement. It always feels like the easiest colors to use to accomplish this look are green and/or white, but you can use any color you have a particular surplus of in your garden.

Extending Vase Life

Once bouquets leave our hands, we want to make sure we give our customers and friends additional ways to increase the vase life of their flowers. To assist with this, we provide customers with a commercially available flower-food packet from Chrysal, a chemical compound that provides nutrients for flowers to thrive.

Aside from these little packets, we've also observed that quite a few well known home remedies actually work really well.

Here's a few other great suggestions to prolong vase life on flowers:

1/2 TEASPOON OF SUGAR DAILY WILL FEED MOST FLOWERS, and 1/2 teaspoon of bleach keeps the water clean. Use both of these together! The bleach keeps the water clear and the sugar gives them something to eat.

CHANGING THE WATER DAILY AND TRIMMING A SMALL SNIP at the base of each flower stem is highly recommended and will extend the blooms' vase life. Most flowers should last for five to 10 days if you are diligent about changing the vase water and re-trimming the ends of the stems.

GARDEN-GATHERED

SUMMER IN A VASE

Garden-inspired designs contain multiple floral varieties with lush, vibrant contrast. These examples appear freshly gathered by hand from the cutting garden and placed into the vase. From early-spring blooms shown above to the early-summer blooms shown below, there are endless options to mimic a garden's wild imperfection.

SHADES + TEXTURES OF AUTUMN

This arrangement designed by Adam captures the essence of autumn. The beautiful compote design includes fall leaves, dahlias, cosmos, black-eyed Susan, ninebark, strawflower, chrysanthemums, hyacinth bean vine, and accents of saliva.

BULBS IN BLOOM

Fritillaria, spring anemone, grape hyacinth, garden roses, and perennial anemone look like a painting in this compote design.

TO HAVE + TO HOLD

A bridal bouquet composed with locally grown blooms of Sweet Love and Sweet Nathalie dahlias, salvia, sedum buds, feverfew, butterfly bush blooms, ammi, cosmos, and mountain mint.

COLOR COMPANIONS

A simple combination of tulips, stock, grape hyacinth, and greenery looks gorgeous in a teal-glazed ceramic vase, accentuating analogous cool tones.

SWEET NOSTALGIA

Using farm-style design and photographed with beautiful, natural light, Jenn creates a simple country bouquet vignette with tulips, hyacinth, ranunculus, and eucalyptus.

ROMANTIC ENCOUNTER

Friends of the farm provided a gorgeous painted backdrop and styled this romantic mix of garden roses, dahlias, cosmos, ammi, scabiosa, and sweet pea greens.

A collector of antique milk glass, Jenn uses soft blush colors including garden roses, dahlias, and mint, with accents of ammi, scabiosa, perennial anemone, and delphinium.

GROWER'S CHOICE

WARM SENTIMENTS

Surrounded by antique frames and wooden boxes, an old crock doubles as a vase for a large centerpiece filled with celosia and dahlias for a farm-to-table dinner.

Late-autumn's glory is reflected in this design with spirea greenery, crabapple branches, rose foliage, arborvitae, sweet Annie, snapdragons, chrysanthemums, celosia, dahlias, and autumn leaves.

HARVEST STILL LIFE

Wedding + Event Florals

If you've grown flowers in the past, you probably already know that it's a pretty natural progression to be asked to provide flowers to family, friends, or even paying customers for baby showers, bridal showers, parties, or weddings. The good news is that you get to decide if this is something you want to do! You can pick and choose the type and size of events for which you want to provide flowers. The best way to do this is to gradually dip your toes into the water and slowly try out some of the options we'll cover in detail below to decide what works for you.

There are several ways to offer flowers for weddings and other large events. Event hosts may choose to grab bouquets at farmers' market, and we've had bridal parties choose bouquets from our offerings to carry down the aisle.

In this case, we do not allow customers to reserve flowers in advance, nor request custom colors for bouquets we bring to market. Everything taken to a farmers' market is what we have seasonally available on the farm, and is offered on a first-come, first-served basis. Bringing bouquets to your farmers' market stall is a wonderful way to share the joy of your flowers with others in a fairly low-stress way, while also giving customers a great opportunity to support local flower farmers.

We also offer customers who want to arrange their own flowers the opportunity to purchase DIY Buckets of Blooms from the farm, available in two options. The first is called Grower's Choice, and it contains an equal mix of seasonally available greenery, focal blooms, and filler flowers from the farm. Each bucket contains around 50 stems, and is sold for $75.

The other DIY Buckets of Blooms option is called Wedding Flowers, and it contains a specific color combination, as specified by the bride. This mix is similar to the above, but is priced just slightly higher at $125. When the couple has specific color requests such as "no yellow, pastels, or fall colors," we suggest they purchase a Wedding Flowers bucket of blooms.

The whole idea around DIY flowers is that the customer is going to design their own wedding flowers. That said, it's sometimes easy to get pulled into additional conversations that may not necessarily enter into DIY territory.

Maybe your goal is to simply provide DIY blooms, as well as make the bouquets and

A flower bar setup for a private wedding party contains the best seasonally available blooms. We often coach wedding parties and teach styling tips as they create their own bouquets.

personal flowers for a smaller wedding, or maybe you want to take on large-scale weddings with all the flowers. As you gradually try things out, you'll find your way and realize what feels comfortable to you. One thing we strongly suggest: Hire additional help for large-scale weddings and other events.

For time estimates, a bridal bouquet typically takes about an hour just for the design, not taking into account the time to cut and prepare the blooms. For a Saturday ceremony, we strongly suggest prepping the blooms on Thursday, and devoting Friday to designing. We also recommend having a large mirror handy in your work area, so that during the design phase you can see what the bouquet you're creating looks like. It's a very useful tool for wedding work!

Trying to figure out how to design a bridal bouquet can take some time, but you can check out our online tutorial on how to make a hand-tied bouquet. Believe in yourself and dedicate time each week to practice floral design.

We invest in ongoing education to learn new wedding design techniques, enhancing the services we provide clients. Shown is a seasonal, nature-inspired "hedgerow," which Jenn created during a floral design workshop. We now use this technique to create living aisles, tablescapes, and large event installations.

Wedding Lessons Learned

We encourage you to build a FAQ document to post on your website or send as an email to share your guidelines with potential DIY Buckets of Blooms customers. Here are some other suggestions:

LIMIT EMAIL COMMUNICATIONS: Because the DIY Buckets of Blooms process should be straightforward, we offer two email exchanges with our customers as part of our contract. If more communication is required, we charge additional fees.

FLOWER SELECTION: Flowers are seasonal, so it's hard to know exactly what will be in bloom on any given day. Yes, we have a general idea, but we have learned to never promise anything, because that's when things don't work out. We steer customers well away from specifying the blooms included in DIY buckets unless we know their wedding date will be around a major bloom time on our farm. DIY flowers are meant to be seasonal and enjoyed as such.

PROVIDE AN ESTIMATE FOR PERSONALS: Customers often ask how far a bucket of 50 stems will go, which is a reasonable question. We tell customers to leave at least one bucket for the bridal bouquet and from there, each bucket would cover at least two bridesmaids, as well as any corsages and boutonnieres. For centerpieces, we often do not estimate this, because it's up to the individual how many stems they include for these.

COLLECT HALF OF PAYMENT UP FRONT: When you send your contract out for a signature, collect one-half of the payment up front. This helps solidify a mutual commitment for the reserved date. We typically collect the balance upon pick up.

DELIVERY VERSUS PICK UP: DIY Buckets of Blooms are 100% pick up at our farm. We've attempted to deliver in the past, but what you'll find is that timelines always fall by the wayside during weddings, and people often run late or forget altogether — which is difficult if you're out on a delivery route and don't have time to wait around. Generally, a specific delivery date and time is set during the initial round of communications, and we have a pretty good idea of when our customers will be picking up their flowers.

FLORAL ASSEMBLY LOCATION: We do not typically provide a farm location for brides and their party to assemble flowers; however, as an add-on service, we do offer our event barn for rent for up to two hours in order to accomplish this, provided it's available.

CONSULTATIONS: Customers occasionally want to meet 1-to-1 to discuss their DIY blooms. We discourage this as much as possible, but do charge a consulting fee in order to meet a customer's request.

FULL-SERVICE WEDDINGS: For those of you who really want to dive into event and wedding florals, offering complete design services is worth further research. Full-service weddings mean that you provide all of the floral elements involved with a wedding day. This could potentially include everything from the bouquets for the bride and bridesmaids, to boutonnieres, corsages, wedding arch, and table centerpieces.

A garden-inspired "living aisle" features dahlias, zinnias, native grasses, and greenery. This 50-foot aisle contained thousands of exquisite flowers from our farm, anchored with a gorgeous wedding arch fit for a secret garden.

A centerpiece in rich and warm-gold florals is accented with fruits and custom name cards.

CHAPTER FIVE

Off to Market

HAVING AN ABUNDANCE OF FLOWERS IS SO WONDERFUL, BUT HOW DO YOU want to enjoy them? There are several ways to appreciate their beauty and the flowers you grow can actually help to fund your operation!

If selling your flowers at the farmers' market, to local florists, or through other channels is something you're interested in doing — even if you're just looking to save up some money for a vacation fund for your family — consider a few of these avenues as a way to generate flower sales.

Farmers' Market

The farmers' market is a wonderful way to sell your flowers, especially as a small and inexpensive way to get started. It's also a great way to meet customers face to face. If selling flowers on a larger scale in the future is appealing, farmers' markets can be the gateway to grow your business.

They are incredibly helpful in getting your name out to the local market via word of mouth, which organically leads to customer inquiries for parties and other events. Once people learn that you're growing flowers, word spreads fast. Farmers' market shoppers typically become regular customers and your most loyal supporters.

We first started selling flowers at our small-town farmers' market in Winterset, Iowa. Our town is considered a rural area, but it's just 30 minutes outside of a major metropolitan area. This agricultural community has a population of around 6,000 people, and the market was the perfect entry point for our business.

Each week, we would bring our farm-grown vegetables and 20 bouquets to the market. The bouquets contained flowers that we usually cut and arranged on market mornings and took us forever to make, because each one was special and unique.

An eye-catching display shows off the best of PepperHarrow Farms' seasonal blooms at the farmers' market. These wrapped takeaway bouquets are customer favorites, popular for their portability, wrapping, and ribbon presentation. Customers often remark that the wrapped bouquets feel like a present, which they buy for themselves or for their friends and loved ones as gifts.

Then we learned over time that it's best to harvest flowers for a Saturday farmers' market on Thursday evenings or Friday mornings, using the remainder of Friday to bundle and wrap bouquets for the market. We also learned that having a repeatable design for our bouquets is critical for saving time in the creative process. Our goal is to keep the production of each bouquet to about two minutes.

> When customers told us they liked to buy our bouquets and arrange them at home, we realized we didn't need to spend extra time on design, since this was something our customers enjoyed doing themselves.

For our farmers' market setup, we initially began with red-and-white gingham tablecloths, which conveyed a perfect country atmosphere. Think lots and lots of burlap, apple bins, antique crates, and other props straight out of the barn. It's good to look back and see the progress we've made since that first season.

For our current setup, we use black tablecloths. A white tent avoids the cast of any other color on our flowers. Both of these provide a neutral backdrop to highlight our gorgeous flowers and wow customers. PepperHarrow Farm branding is also very important for our setup and presentation, offering a consistent look and feel with black and white ribbon on our wrapped bouquets and our black-and-white logo on bouquet labels and signage. With the table setup and tent, all of these elements convey a visually pleasing experience for shoppers who enter our space — and let's not forget the stunning flowers that steal the show anyway.

Another Way is CSA

Community Supported Agriculture (CSA) is a great way to start with a small group of flower customers by providing bouquets on a weekly or bi-weekly basis. The wonderful thing about CSAs is that they provide consistent income. When you sell subscriptions, you collect money upfront to pay for your investment into seeds, plugs, and other supplies.

If you've never tried to operate a CSA before, it's wise to start small with the number of CSA spots you offer. A good number to target when first starting out is five or fewer spots per month, offering either a bi-weekly or monthly subscription. As you get comfortable with the process of growing flowers and are consistently harvesting enough blooms to cover the time span of your CSA program, you can increase the number of spots you offer from there.

Being conservative and learning a lot — while offering just a few bouquets each month to start — is a great idea for at least the first two years. Managing too many bouquet orders can be stressful if your garden isn't producing enough flowers to cover the spots you've sold, or isn't producing enough blooms to cover the span of time you've committed to your customers.

Committing to a specific number of bouquets on a regular basis will teach you a lot about how many flowers and bouquets you're able to produce, given your growing space, as well as help build your confidence with how the process works. We didn't start offering CSA spots until our ninth year, although we had been asked by multiple customers for years. We postponed a CSA program because we were concerned about having enough blooms to fulfill bouquet orders each week. In hindsight, we probably had the amount of flowers to support a CSA in our sixth year!

In 2020, when we learned that our local farmers' market was canceled for the year, we scrambled to find other channels to sell our flowers and keep our customers happy. Luckily, we had a bit of time to come up with ideas on how to shift our business model to survive. Launching a CSA helped us offset some of the loss of revenue from the farmers' market, which helped keep our small business running.

> Unlike most CSA models — where customers typically pick up their flowers at an urban location — we only offered the option of "on-farm pick up" for our bouquets.

We did this to give our customers an opportunity to get out of the house, out of the city, and into the open air to enjoy the abundant flower gardens at PepperHarrow Farm. Spots were limited, so we could offer social distancing, and a private, exclusive opportunity for our customers.

It was a great experience for everyone. We loved getting to know our customers on a more personal level. Their visits led to more interactive and intimate conversations than what we typically had time for at the farmers' market. Our customers loved getting to know us as well. When they came out to pick up their bouquets each month, customers saw our hard work to care for the farm and they loved seeing the transitions of flowers throughout the growing season. It was pretty dramatic for them to see sunflowers growing one month, and a field of wildflowers or cover crops upon their return.

With the growing popularity and overall success of our CSA program, going forward we plan to expand our CSA offerings to additional customers. To add a bit of risk and to experiment with market interest, we're also going to try out a few new types of offerings we've never tried in the past.

CSA Offerings

Our CSA program consists of several offerings, which sometimes vary from year to year, but here are some of our regular ones:

MONTHLY BOUQUET: Customers receive a fresh bouquet of flowers in a vase, which they pick up on the second Saturday of each month from May through September. Customers are instructed to bring their vase back for use the following month, but they keep their vase after the last pick up. This is our most popular CSA option.

MONTHLY YOU-PICK BUCKET OF BLOOMS: Customers come to the farm to cut their own fresh flowers on the second Saturday of each month from May through September. This option is popular with local artists, including painters and photographers, as well as with people who want to practice their floral design skills.

WEEKLY BOUQUET (CHOSEN BY MONTH): Customers receive a fresh bouquet of flowers from the farm in a vase for pick up each Saturday for a selected month (May through September).

Spring Seasonal CSAs

POLLINATOR PLANT SHARE: Consists of a deep-cell plant tray of 38 mostly native plants that are pollinator-friendly, designed to help customers start their first pollinator garden. This is offered for farm pick up, usually early in May.

CUT-FLOWER GARDEN PLANT SHARE: Consists of 30 small plant plugs of our best flowers for cutting. This is a great option for customers looking to build their own backyard cutting garden. We include a mix of 80% annual and 20% perennial plants in the share, so some of the plants will come back the following year. This is one of our most-popular CSA Shares in the spring.

DAHLIA TUBER SHARE: We package a mix of 15 of some of the best dahlia tuber varieties that we grow on the farm. This is offered for farm pick up early in May.

HERB PLANT SHARE: Contains a variety of 20 different types of herbs to establish a home herb garden. This collection is perfect for people who love to cook and garden.

PEONY BOUQUET SHARE: Contains 10 stems each week for three weeks of peony season. We offer this in early spring for May pick-up on the farm. The bouquets come with a vase.

SPECIALTY CUT-FLOWER PLANT COLLECTIONS: You can do this with popular flowers that may be difficult for your customer to find at local nurseries. We like to focus on lisianthus and hellebores specifically, because they're staples on the farm and are hard to find locally.

Fall Seasonal CSA

FALL DAHLIA BLOOMS: Customers receive a gorgeous mix of 10 stems of our favorite dahlia varieties on the farm. With over 10,000 dahlia plants blooming on the farm, and hundreds of stems harvested daily, each week is sure to be a different mix of stems.

Offering You-Pick

Do you have enough flowers to allow others to cut them from your garden? You might want to consider a you-pick program. Customers love cutting their own flowers, because it gives them a chance to connect with nature and hand-select the special varieties they want to include in a bouquet. This is also a great way to reduce labor costs for a flower farmer, because there's no need to put together a mixed bouquet. The customer essentially does all the work for you.

You-pick is another one of our newer services that we developed to enhance our customers' experience with PepperHarrow Farm. There are two options available: The you-cut bouquet of blooms consists of around 25 stems of flowers, while the you-cut bucket of blooms consists of around 50 stems. You-pick is scheduled on Saturdays over a four-hour period.

Customers purchase a spot online for either one of services, during the week prior to the Saturday they want to visit. We provide a vase for the bouquet, either a wrapped

Expansive flower fields and copious locations for photo opportunities are all part of the amazing hands-on experience provided by the you-cut flower bouquet opportunity. Watching children's curiosity as they cut and learn flower names with the help of their parents is one of the most rewarding parts.

pint jar or one finished with a ribbon and sticker. We also provide a florist bucket for customers who purchase buckets of blooms. Our customers just bring their own floral shears, or we offer them for sale if they needed a pair.

> By using our website to sell preordered tickets, we always know exactly how many customers to expect each Saturday.

Selling you-pick spots in advance was a huge advantage for us. We knew exactly how many vases to prep and how many buckets to have ready; it also ensured that our blooms weren't being over-cut.

You-pick has been incredibly successful for us and has given our customers a chance to see the flower farm in person. Of course, we're always worried we won't have enough flowers, and that is a legitimate concern early in the spring. But once the season takes off around mid-June, there's plenty of flowers for the number of spots we open up each week, which is usually 25 people for bouquets and 10 people for buckets.

If you offer a you-cut experience, it's best to keep standard, regular days and hours as available times for your customers. Customers need reliability and consistency to be able to follow through with purchasing.

Other Opportunities

Selling bouquets or arrangements on Facebook, to local florists, through pop-up shops, and at local artisan or craft markets are also great ways to generate sales, expand local interest, and create demand for your flowers.

Additionally, front-porch bouquet sales and flower stands have become increasingly popular avenues for small-scale growers to sell their flowers. People enjoy the accessibility of buying local flowers from the front porch or flower stand, and it's a fun, authentic way to reach and connect with customers.

Floral enthusiasts also love opportunities to get into the garden to have a unique, wonderful experience. Consider sharing immersive garden events, such as floral-design classes, floral retreats, yoga in the garden, garden dinners, and garden-related workshops like our pollinator education series. The possibilities of what you can offer are plentiful.

If hosting events is something you're interested in, the sky's the limit as to what you can provide. Think outside the box and get feedback from your friends to generate ideas. It's also great to connect with local influencers, or other local creatives to find ways to collaborate.

How to Price Flowers

Even before you decide on an outlet to sell your flowers, it's important to think about how much you might charge for them. Markets and pricing can vary depending on location — think urban versus rural.

In general, urban areas offer a larger body of potential customers, as well as customers with disposable income. At the same time, rural areas are trending, fast-growing communities, and you'll find wonderful supporters of local business. Rural locales provide great opportunities to develop brand trust and community.

It can be a little intimidating trying to figure out how to price bouquets and arrangements, but we have a few tips to share as a rough guide on pricing your flowers and wholesale bouquets.

Keep in mind; however, if you're selling flowers directly to customers, the retail price you charge is usually double what is shown below. Remember, market pricing varies greatly by region and you should be flexible, and adjust prices accordingly.

START by doing a little market research in areas where you hope to sell your bouquets. Observe what existing businesses may already be selling. Also, consider purchasing flowers from a local grocery store.

DECONSTRUCT and analyze the individual flower stems within that bouquet. Focal flowers are more expensive, followed by filler ingredients, with foliage being the least expensive. Attempt to calculate each stem versus the price you paid for the bouquet.

KEEP IN MIND the wholesale price for bouquets to grocery stores is typically 50%, or one half of the retail price. This is always negotiable, but it's a pretty good rule of thumb. As you're deconstructing a bouquet from a local grocer, stem by stem, be thinking about this in the back of your mind.

Many people who sell flowers leverage Boston Terminal Pricing, which is a valuable resource found via a quick Internet search. Because prices vary so much by region, this report can only be used as a rough guide.

Here's a little secret: Become a customer of a local flower distributor in your region. This will provide a better guide to regional market pricing for your flowers.

Left: A monarch butterfly enjoys taking a rest and replenishing its energy on a dahlia bloom.
Right: One of many bee species on the farm enjoying an afternoon of foraging on sunflower pollen.

Creating a Pollinator Habitat

Set aside a section of flowers as a pollinator habitat. In doing so, you'll not only be giving back a small portion of your garden to nature, but you can use it to provide an educational opportunity for children. You can do this in even the smallest of spaces. It's simple to devote a 5-foot-by-5-foot parcel just for this purpose. A nice pollinator seed mix can yield a pretty wildflower garden that not only looks lovely, but it will help feed and provide habitat for our winged friends and beneficial insects.

We recently added a small pollinator garden on the farm and included a placard with a QR code, so visitors can scan and see more information about the nectar-providing plants it contains. We love offering interactive and educational experiences for visitors to the farm. Of course, if you plant a pollinator garden in your home garden or property, you'll have to name it! That's the best part.

For more information about pollinator plants for your area, visit www.pollinator.org to find suggestions and lists.

Sharing the Beauty

If your garden is exploding with blooms, there are many ways to share your floral bounty and bring joy to other people's lives. Sprinkle a little flower joy! Take the time to slow down and appreciate your flowers. Never miss an opportunity to cut flowers to simply appreciate, enjoy, and share with others.

RANDOM ACTS OF FLOWER KINDNESS: Have you ever thought about just leaving a bouquet in a random place with a "take me" sign and uplifting note? Or, maybe gifting a bouquet to a hardworking educator or teacher? Little acts of kindness like this can start a ripple of positivity that eventually creates a wave.

PHOTO FINISH: Capturing vivid, bright colors and the sheer beauty of flowers through floral photography is a great way to enjoy the diversity of the flora you've cultivated in the garden. It's a gift not only to revel in, but also to share with others through social media, prints, cards for family and friends, and photo albums.

Pulling on this artistic thread is a natural progression, and it ties in well with growing flowers. The captivating, ephemeral beauty of the garden through photography is a perfect outlet, but painting is another soul-fulfilling way to appreciate nature's goodness.

WATERCOLOR CLASSES: We offer classes at the farm each growing season. These classes are a chance for novice painters and pros alike to relax, appreciate the beauty of flowers, observe their intricate details, and create a unique masterpiece. You can also invite outdoor artist groups to come to your garden or property to paint. We did this for the first time recently and what a beautiful scene it created. Not only was it fun watching the artists as they recreated our flowers on paper, but the amazing amount of talent we saw with the finished paintings was incredible.

DRYING BOTANICALS: You can also investigate pressing or drying botanical ingredients as another form of artistic expression through the flowers you grow. You can make wreaths, or create other floral-art mediums with your dried flowers. Press and dry flowers throughout the season and store them for use in fall and winter as fun, downtime activities. Drying flowers may seem like a lot of work during the height of your growing season, but we always take a little time out each week to set aside and press a few of our flowers for off-season enjoyment.

SETTING A BEAUTIFUL TABLE: Using locally grown flowers is such a special way to celebrate any occasion. Pretty, edible floral touches — such as lavender lemonade with lavender grown right from the garden — make special occasions even more special.

Maybe even consider getting friends together to help make centerpieces to decorate for a dinner party, charitable event, wedding ceremony or reception, host/hostess gifts, residents of a local care center, or just to brighten a stranger's day.

Children's Lessons from the Garden

There's so much you can teach children through gardening. From seed starting and weeding to tending flowers, there are wonderful life skills hidden within each step of a plant's life-cycle, and each lesson can become a foundation for success later in life.

Patience, commitment, responsibility, and work ethic are among the top skills children can learn through gardening. Watching a tiny seed grow into a flower, tending to it daily, and waiting patiently for the magic to happen is fun for kids.

Although not as fun, it's also important to teach children that not everything is successful. Sometimes seeds fail to germinate and grow, or deer eat all the tulips and sunflowers. Part of growing a garden is about accepting a little bit of failure, and in some cases, learning from your failures to become better next time.

> It's great to encourage youngsters' participation in the garden, even by helping with watering, deadheading flowers, or the dreaded chore of weeding.

Even though these things might be a little boring for kids at times, it's important to show them that simple tasks can pay off with a reward of flowers in bloom. Adding a little excitement, and talking about the garden at the dinner table each night is a nice way to increase participation and anticipation of the blooms to come.

Learning how to garden gives children a great lifelong skill set that will serve them well. Understanding they can grow their own food to nourish their bodies is an important lesson to pass along from one generation to the next.

Children always learn best from hands-on activity, so give them a dedicated garden spot to grow their own plants. Have them plant easy-growing, bright flowers such as marigolds, sunflowers, and zinnias. As the flowers grow, children can tend to them and be accountable for their own little garden spot and watch it flourish.

Some of our fondest memories have been spent in the garden with our children. We loved learning about their favorite flowers and why they like them. Although it doesn't happen as much now that they're teenagers, we always enjoyed sending them off to cut flowers to make their own bouquets. As with typical design classes, each bouquet looks as different as its maker, because it's an extension of their own artistic expression.

Now that they are older, our three children view living on a flower farm as a way to make a little extra money to pay for gas and other items they want, which means they often ask if they can help us out. They've also found it pretty useful to live on a flower farm, because they have us make corsages and boutonnieres for them and their dates at homecoming and prom, as well as special bouquets for their own dates, and flowers for favorite teachers.

We love to teach our children about growing flowers and instill in them an entrepreneurial spirit.

Growing flowers to sell at the market has been an opportunity for our kids to observe this, and they've been inspired to try some of their own ideas. Each year at the farmers' market, they would try to come up with something to sell that was theirs alone, such as face-painting, one-of-a-kind drawings, and of course, their own bouquets to sell at market. When someone picked one of their bouquets, the joy it brought was a priceless feeling!

Our youngest daughter dreamed up a kids' watercolor class on the farm. She taught kids (with parent chaperones) how to do watercolor art, and instructed them to paint their favorite flower. It was a smashing success!

She deposited half of her proceeds into her college savings account, and donated the other half to our local animal shelter. We are so proud of her concept, as it allowed her to express her artistic ability, as well as develop teaching and leadership skills. It was also wonderful for the kids who attended to not only experience the farm, but to see one of their peers take the lead.

This year, she came up with the idea of having a flower stand to sell extra bouquets of flowers during onsite events. Through this idea, she's taken personal accountability to spend extra time in the garden weeding the flowers she's growing for bouquets, budgeting costs she's going incur while producing her bouquets, and forecasting sales and net profit. She has big plans, folks!

Outside of raising our own children on a flower farm, we love seeing families and we invite local children's groups out to the farm to learn about what we do. We've hosted a variety of groups from the Girl Scouts to home-school family networks. Educating kids about farming, growing flowers and food, and making it a fun experience for them gives a foundation they can either act on, or draw from later in life. It's so important to share farming with our youth and encourage them to get growing.

The activities we offer vary by group and by the season, but we especially love the fall because we usually have extra pumpkins to share. Letting children paint pumpkins is a little messy, but just the best activity! We've also given young visitors the opportunity to cut their own bouquet of flowers.

Running barefoot through our lavender field, our daughter enjoys frolicking in her backyard playground.

CHAPTER SIX

The Dormant Season

Once the growing season is over, it's time for cleanup, for stepping back from the hustle of growing flowers, and shifting to our end-of-season activities. This is a great opportunity to catch up on tasks and to-do lists that have been pushed to the side while farming. Dormant season also provides a chance to evaluate, take some time for perspective, be creative, and come up with strategic plans for the coming growing seasons and beyond.

It's important to note that this time shouldn't all be spent working on cleaning up and strategic planning. Downtime allows you to regroup, gather yourself, invigorate your creativity, and maybe even read a book or two. It's a critical period for recharging your spirit while relaxing and appreciating the quiet of winter. Take advantage of your slow time and you'll soon find yourself anticipating the following season and all it promises.

Winter Preparation

At the end of the season, the flowers and vegetables in the raised beds are cut to the ground, with the spent foliage thrown into the compost pile to break down into wonderful organic material for the following season. However, when it comes to most of the larger production areas, we leave most of the material in place for a couple reasons.

We found that leaving the plant material and any row covering in place provides habitat for any birds that arrive in winter. They seek refuge, build nests, and forage for leftover seeds in these garden beds during the winter months, and this turns into a favorite spot for them. Keeping the row covering in place also prevents our precious topsoil from blowing away during windy, winter months.

This time is also spent spreading additional mulch around the garden roses, and topping off the roses and other mulched areas with a snug cover of wood chips to provide further winter protection. In the process, we've found it keeps the farm tidy, and takes a task off of the spring to-do list.

Dahlias are Different

During the middle to later part of fall, we begin plant-tagging our dahlias, writing each variety name on a tag, and securing it around the stalk of the plant. Even though our detailed maps track the dahlias grown in each row, the process of tagging dahlias when we dig them is especially helpful after frost hits and it's hard to tell the color of the dahlias.

Occasionally, we have a random color growing in a row where it doesn't belong. We take the time to pull these out before dahlia digging, otherwise the rogue color will likely get mixed in with tubers being divided. This is an especially important activity if you plan on selling dahlia tubers, as you want to ensure correct labeling so customers receive the particular dahlia variety they ordered.

In late fall, "The Great Dahlia Dig" is a huge activity on the flower farm. Because we're located in a part of the country with cold winters, we dig and store our dahlia tubers each year.

Crates of freshly washed dahlia tubers are ready to transport to the air-drying area.
After a few days, they are dry and ready to be divided and stored for winter.

With roughly 12,000 dahlia tubers, this task usually takes four people over two long days of work. We like to start digging the tubers as early as possible, sometimes even mowing down the plants and pulling them from the ground. It's always great if we have a mild, dry fall to tackle this large farm task. If we wait too long, it usually gets wet and cold, which means conditions are miserable to work in.

WASHING THE TUBERS: Once the dahlias are dug out of the ground — weather permitting — we set up a washing station to give them a rinse, and then allow them to dry for 24 hours. We wash the tubers thoroughly, removing as much of the soil and earthworms as we can. If the weather is too cold, we usually bring the dahlia tubers into a heated work area, keeping them in their dirt clumps until we have time to wash and divide them.

Allowing the dahlia tubers to dry off for a few days after the initial wash is a critical step! If you store your tubers with any moisture, they will mold during storage and it will spread rampantly.

DIVIDING THE DAHLIAS: After the dahlias are completely dry, dividing the tubers is the next task. This project is a constant focus for us from October through December. We typically spend at least three long days each week focused on this activity alone. It's a ton of work, but it's also a great way to catch up on podcasts.

We like to divide before storage, because it limits the amount of molding that may occur during storage. It always seems like we find more soil or earthworms hiding in the middle of our dahlia clumps when we divide. This is especially the case on dahlia tubers that grow in big, tight clumps. Even if you don't want to fully divide all of your tubers, we recommend splitting the dahlia tuber clump in two for storage.

STORING THE DAHLIAS: Once the dahlias tubers are divided or split in half, we store them in kraft paper-lined crates, surrounding them with peat moss, and then we place them in our walk-in cooler. The cooler is kept between 40-65 degrees F, with 80-90% humidity throughout the winter. We have a heater, humidifier, and fan going at all times, as well as a digital temperature stick thermostat to monitor everything for us. The heater has an automatic shut off when it hits a certain temperature, which is critical to make sure it doesn't get too warm in the space. The digital temperature stick has been an invaluable tool for us and for monitoring our dahlias throughout winter. It gives us an up-to-the-moment look at temperature and humidity, and alerts us when either of these is above or below the thresholds we have set.

Don't let our use of a walk-in cooler deter you from this process. For a home gardener, any space will work as long as it can be maintained with the temperature and humidity requirements we've outlined. Many gardeners even simply keep their dahlias in their clumps of soil, throw them in a box in a storage area, and have success as well. This process is something you should experiment with to determine what works for you.

Cleaning Tools, Tidying Workspaces

Cleaning seed-starting areas, trays, and pots is an arduous activity, but one that is certainly critical to providing a safe, sterile growing environment when seed starting resumes in late-January and February.

Seed-starting supplies, including plug flats, propagation trays, planting trays and buckets, all get a good wash first with a bit of soapy water, then a quick dip in water with a bit of bleach. We've found this is the best way to prevent pests, such as fungus gnats or aphids, from appearing in the growing areas the following season. Both of these pests can totally decimate an indoor growing operation, especially during early-spring months when the greenhouses are totally closed up.

> It's definitely worth the time you invest to make sure your growing supplies are clean and sterilized. We also spend this time cleaning and sharpening our tools.

By the end of the season, most of our floral snips are a bit dull and definitely need cleaning. They're first brought inside and are washed by hand with a soapy washrag, and well rubbed with a steel scrubber; then we dry them and bring them out to the shop for a good sharpening. The washing process is also repeated with any other tools we used during the season, such as rakes, shovels, and planting tools.

Inevitably at the end of the season, we usually have to go around the farm to pick up random tools left in different areas. Most often, a shovel or two were left out in the fields, or in one of our barns. Grabbing all of the tools and returning them to a central location helps us feel quite organized when the new season begins.

The entire workspace where we divide dahlias, as well as the main production area where we make bouquets, gets a good tidying in late winter. Vases are organized and staged by color, the floors are swept, and excess waste is removed. All of this is done in an effort to prepare for spring's arrival and most importantly, the massive early-spring shipping of our dahlia tubers around the country.

Prepping + Planning

The days that aren't spent dividing dahlias are spent on business planning. This is an all-encompassing task, but it serves us well once the growing season takes off.

CSA EVALUATION: We typically start by evaluating our CSA offerings and events for the following season. We discuss which CSAs programs we want to keep, eliminate, or introduce. This involves evaluating profitability for each of the previous season's CSA products, brainstorming ways to improve the programs, and discussing how many spots to offer.

After CSA evaluation is completed, we take a look at our onsite events with the same level of scrutiny. Once the business analysis is complete, we secure calendar dates for the onsite classes we offer such as: seed starting, floral design, and container gardening.

COLLABORATION WITH OTHERS: We enjoy collaborating with other small businesses to create fun and compatible experiences on the flower farm and strategize about product placement, pop-ups, and schedules. This step can sometimes take quite a bit of time, but it's fun to get together with other local business owners after a long season of farming. The camaraderie is priceless, and we often leave these meetings inspired to try new ideas, and ready to tackle ambitious projects for the following year. We can't stress this enough: find ways to collaborate with other local businesses. It's amazing how much you'll be able to support each other.

REVIEW OUR NOTES: During the height of the growing season, we record a lot of observations including making notes about the varieties of flowers we loved growing, ones that grew well with little maintenance, and those that bloomed like crazy. We also note special varieties we saw other gardeners growing for possibly adding to our own beds and borders. It's a good idea throughout the growing season to keep quick notes on your cell phone or take screenshots to reference later — even when you're busy and don't have time to sit down to document it on the computer.

Once the seed catalogs start arriving, you should have a great list of things you already know that grow well in your garden, and you may also have a pretty good idea of what things you'd like to add the following growing season.

ORDER SEEDS AND PLUGS: Each year, we begin our seed orders with standard seeds we grow each and every year, but we always include about five to 10 new flowers that we've never tried before. We may or may not be successful with them, but these fun additions bring us happiness and excitement for the upcoming growing season. Looking back to our previous year's list of new varieties we tried, it's encouraging to see how our experimentation led to some great successes. Examples include lavender, pansies, smilax, bush clematis, and pimpinella (also called burnet-saxifrage).

Aside from seeds, we also research which flower plugs are available. Plugs are little plant starts that give growers a jumpstart on the season. We usually order multiple lisianthus varieties, which are annuals, preferring to spread them across three different shipping dates to make sure we succession plant them. We also order perennial plugs, which are the plants we want to add as permanent additions to the farmscape. Almost all of our lavender plants originated as plugs.

Although less commonly ordered, we love sourcing and planting perennial plugs, because they're a reliable investment with consistent blooms each year.

Purchasing new dahlia varieties is another wintertime activity. It's fun to research and add unique varieties to our list for the coming year. We look for dahlias that we think our customers will love to design with, and sometimes even seek out their opinions. Trust us, if you ask a florist to request a preferred dahlia palette, you'll likely receive one pretty strong color message: blush, white, and burgundy. These are the dahlias most retail florists need for weddings and other events. We also select new dahlia varieties that catch our eyes for use in our own designs or market bouquets.

> After the list of must have dahlia tubers is completed, we plan how we're going to procure all of them, and from which vendors we'll source what we need.

Some are ordered from distributors, but the lion's share of the varieties we grow here are sourced from other flower farms across the U.S. The quality of the tubers from most flower farms is superior, and we love supporting other small growers.

During late fall and into winter, we also make shopping lists of supplies we will need the following season. This includes everything from seed-starting medium to vessels for design classes. It's great to order all of these items at the beginning of the season, so we're prepared before we get busy growing flowers. It's easy to postpone these tasks, but we never regret taking care of things early, and benefit from crossing them off the list while there's time.

Marketing

During the winter months, we spend a fair amount of time updating our website with new pictures from the growing season and refreshed verbiage. This is a great way to keep any website fresh and relevant. It's also necessary from a marketing standpoint to make sure target customers can find you on the web. We also review and strategize about possible blog posts that will support our marketing strategies.

Blog posts can be anything that you want to share about your farm, from details about growing practices to favorite flowers. We've found blog posts are a great way to share a more in-depth look at either the day-to-day activities, or banner celebrations and events that are held at the farm.

A blog can share beautiful visuals, as well as descriptive writing. We like writing blogs, because it's an artistic way to tell our stories, and show pictures of the farm we love.

One of our favorite late-evening activities during farming season is shooting photos of our flowers and floral designs. The sunsets provide beautiful backlighting against our farm's vast vistas, as we capture images of the beauty and magic of the blooms.

The same thought process is what we implement when we plan for the upcoming season's social media content. While not everything is planned out for the following season (because it's also important to be real and live in the moment), we like to have an idea of special dates where we use specific images to help generate customer interest and participation.

We take photos of flowers all season long. Some of them make it into social media posts, and some we save for specific marketing projects, or simply don't make the cut during the height of our growing season. Those bonus images are usually a smashing success when we post them in the middle of winter.

The best way to build a nice library of interesting social media content is to take as many photos as you can of your flowers, while they are in various stages. It has been an invaluable marketing tool to have lots of beautiful images to pull from for blogs and social media posts. Not only do we get to share small glimpses of what we love with our community, we get to add so much beauty with the world.

Adam was inspired to create a large installation of fall foliage and branches for an open house on the farm. It served as a beautiful backdrop against the red barn, greeting guests as they arrived. A bonus: The setting was photo-ready for our visitors' self-portraits.

Taking Time for Creativity

Growing flowers and running a small enterprise is very hard work, but what's hard work without a reward? We are big believers in devoting time to playing, creating, and finding inspiration. The best part of the winter months for us is taking a break from the hustle and bustle of our busy season. During the growing season, we work really hard and tend to burn the candle at both ends, so during the winter months, we spend lots of time rejuvenating.

As often as possible during our growing season — but mostly during fall, winter, and early spring — we spend time walking the nature trails around our property, or walking at local parks and nature preserves.

There's so much inspiration hidden within the simple but beautiful perfection of the natural world.

Time away from the farm is a necessity. We devote our end-of-season downtime (late-December through January) to escape for a family holiday. It's usually nothing fancy, and sometimes as simple as spending a night or two in a large city a few hours away, but we schedule these special moments to reconnect and have a little fun.

We also try to escape the farm at least once during the middle of the growing season. This mini-getaway breathes new life into our spirits, and gives us a new perspective when we return to the farm. Creatively speaking, the places we visit during our great escapes are usually agricultural communities that inspire us with new concepts to incorporate into our business, or spark some ideas for us to pursue and evaluate during the winter months.

On the farm, we escape to the small flower cottage, which we converted into an art studio oasis away from the main house. It's a quiet space devoted to tapping into the creative part of ourselves that doesn't often experience unscheduled time. The cottage is furnished with tables and benches, a large easel, and art supplies.

Winter is also the ideal time to put together our vision boards for the following season. We start by clipping inspirational or directional images, or keywords from magazines. Once all of our favorite images and keywords are clipped, we begin to piece them together on a large poster board, gluing them into place to create a vibrant, visual, and verbal collage.

The vision board channels positive images and thoughts, which can help your vision manifest itself in subtle or subliminal way. Our board hangs in our office, so we can look at it while we work on our laptops, and continually reminds us of our vision for the next season on the farm.

Resources

PEPPERHARROW FARM
pepperharrowfarm.com

Our online farm shop, features a collection of our favorite tools, seeds, and other thoughtful garden-related gifts, plus lavender products made on the farm.

Instagram: pepperharrow_

Facebook: PepperHarrow

TikTok: @pepperharrow

BIG DREAMS MEMBERSHIP
patreon.com/pepperharrow

Our learning community provides dynamic resources for flower farmers, garden lovers, and farming enthusiasts to come together to learn and grow. Flower-farming educational offerings range from a private online membership community, to exclusive online educational content, and 1-to-1 consulting sessions. You're sure to find a place to call home with our support and online community.

FLOWER FARMING BOOKS

"The Flower Farmer: An Organic Grower's Guide to Raising and Selling Cut Flowers, 2nd Edition"
by Lynn Byczynski

"Cool Flowers: How to Grow and Enjoy Long-Blooming Hardy Annual Flowers Using Cool Weather Techniques"
by Lisa Mason-Ziegler

"Postharvest Handling of Cut Flowers and Greens: A Practical Guide for Commercial Growers, Wholesalers, and Retailers"
by John Dole, Robert Stamps, Alicain Carlson, Iftikhar Ahmad, Lane Greer

GARDEN DESIGN BOOKS

"New Naturalism: Designing and Planting a Resilient, Ecologically Vibrant Home Garden"
by Kelly Norris

"The Layered Garden"
by David Culp

FLORAL DESIGN MASTERCLASSES

Philippa Craddock
www.philippacraddock.com

FLORAL DESIGN BOOKS

"Color Me Floral: Techniques for Creating Stunning Monochromatic Arrangements for Every Season"
by Kiana Underwood

"Fresh Flower Arranging: Step-by-Step Designs for Home, Weddings, and Gifts"
by Mark Welford and Stephen Wicks

HARVEST AND POST-HARVEST SUPPLIES

CHRYSAL
chrysalflowerfood.com
Shop for post-harvest care including our top picks RosePro Hydration.

FLORAL SUPPLY SYNDICATE
fss.com
Open to the wholesale trade only

WHAT WE WISH WE KNEW BEFORE WE STARTED A FARM

12 minutes | 32,000+ views

Hear what we wish we knew before we began! We discuss lessons from our journey and share frank advice to help you be successful. Learn what to grow first and set up your flower farm for success.

LIFE ON A FLOWER FARM

17 minutes | 28,000+ views

Get a look behind all the pretty Instagram posts. Join a typical day on PepperHarrow Farm -- from washing buckets to the dahlia harvest. We design for a wedding and our CSA bouquets, show off our lavender-drying process, and plant cover-crops.

MAKE 100K SELLING CUT FLOWERS?

2 minutes | 22,000+ views

Can you make $100k selling cut flowers? We show you our plan for earning six figures on a flower farm. We discuss sales channels you can pursue in your floral business and break it down to help you achieve $100k in gross sales.

— supplies goods for floral design including floral adhesive, chicken wire, vases, waterproof tape, pin frogs, and ribbon.

COOL BOT
storeitcold.com
Offers DIY and prefabricated cooling units.

JAMALI GARDEN
jamaligarden.com
A great source for vases, garden supplies and floral design tools.

MAYARTS
mayarts.com
Offers a large selection of colors and sizes of quality ribbon.

ULINE
uline.com
Offers kraft paper, tissue paper, rubber bands, and buckets.

SOIL TESTING

AGSOURCE LAB
lawnandgarden.vas.com

WAYPOINT ANALYTICAL
waypointanalytical.com

TOOLS AND SUPPLIES

A.M. LEONARD
amleo.com

BOTANICAL INTERESTS
botanicalinterests.com

FARMTEK
farmtek.com
Offers greenhouses, cold frames, and other farming supplies and structures.

GARDENER'S SUPPLY
gardeners.com

GROWING SOLUTIONS
growingsolutions.com
Offers compost tea systems for growers and gardeners.

NOLT'S MIDWEST PRODUCE SUPPLIES
noltsproducesupplies.net
Offers Whirlybird sprinklers.

SEEDS

JOHNNY'S SELECTED SEEDS
johnnyseeds.com/tools-supplies

RENEE'S GARDEN
reneesgarden.com

BAKER CREEK
rareseeds.com

BULBS

BRECKS
brecks.com

DUTCH GROWN
dutchgrown.com

DAHLIAS

PEPPERHARROW
pepperharrow.com

SUMMER DREAMS FARM
summerdreamsfarm.com

SWAN ISLAND DAHLIAS
dahlias.com

SUCCESSION PLANTING MADE EASY

14 minutes | 14,000+ views

Succession planting is a way to make sure your farm has continual blooms, all season long. We simplify succession planting and share our strategy from seed to bloom -- so you can create your own system to maximize flower production.

FLOWER-FARMING TIPS FOR THE HOME GARDEN

12 minutes | 5,000+ views

Borrow the best advice from the pros. We share flower-farming advice for the home cutting garden. We cover seed starting and spacing and suggest the best succession planting and harvesting methods to yield abundant blooms for arrangements and bouquets.

HOW TO WRAP A BOUQUET

2 minutes | 23,000+ views

Adam demonstrates our favorite method for wrapping bouquets with a full and abundant appearance. This simple, yet effective method is ideal for farmers' market bouquets. It's also fun to wrap flowers for gift-giving.

On Saturday, March 5, 2022, just days before Small Farm, Big Dreams was headed to the printer, PepperHarrow Farm was hit by a deadly EF4 tornado that raced through a 14-mile swath of Madison County, across the flower farm, destroying the shop, barn, flower cottage, event facility and greenhouses. Wind speeds were estimated to have reached up to 165 miles per hour.

EPILOGUE

A New Dream

Having our flower farm hit by a tornado still feels a bit surreal. In the midst of the sadness, we are embracing the future potential of working with a literal "blank slate" as we reestablish our farming operations. All of our existing infrastructure was swept away. The event barn, lavender distillery (red barn), and greenhouses are all gone, but the land remains. We will farm again.

We wanted to share four important highlights from this profound experience:

GRATITUDE. We are very grateful that our lives, our children's lives and our family home were spared. Our dogs are safe and, in the days following the tornado, most of our cats have returned (may we have as many lives as they have)!

COMMUNITY. Very closely connected to gratitude is the profound gift of our community, near and far. The flower-farming community has flooded us with love and encouragement, from growers across the U.S. and abroad. There are those who offered us hard-to-find dahlia tubers from their personal stock, shipped farm supplies, established plant funds, and started flower seeds. Local flower farmers simply showed up to share sweat equity and help with cleanup in the early days following the tornado.

RESILIENCE. Farmers are resilient. In the early days of our post-tornado cleanup, amid mess and disaster, we saw the new green growth of our daffodils and irises poking out of the soil. We take strength from the resilience of our plants and the promise of their blooms.

HOPE. We have 11 years of flower-growing experience under our belts. We know exactly what we need to be successful and how to make those things happen. Now, we get to build a new dream and turn our ideas into a flourishing reality. That hope and deep resolve is propelling us forward into the near future when once again our beautiful farmland will be covered in blooms.

In these pages of "Small Farm, Big Dreams," we shared our best advice to help you turn your flower-growing passion into a successful floral enterprise. Pulling from this well of knowledge, along with our unstoppable spirit and desire to grow beautiful flowers, we're pursuing our biggest dreams yet. We can't wait to begin our new chapter at PepperHarrow Farm. As we propel forward into our next phase, we do so with hope and gratitude. We have big dreams. We invite you to follow along as we begin to rebuild our farm and once again fill our fields with flowers.